Amory Prescott Folwell

Sewerage

The Designing, Construction, and Maintenance of Sewerage Systems. Second Edition

Amory Prescott Folwell

Sewerage
The Designing, Construction, and Maintenance of Sewerage Systems. Second Edition

ISBN/EAN: 9783337034238

Printed in Europe, USA, Canada, Australia, Japan

Cover: Foto ©berggeist007 / pixelio.de

More available books at **www.hansebooks.com**

SEWERAGE.

THE DESIGNING, CONSTRUCTION, AND MAINTENANCE

OF

SEWERAGE SYSTEMS.

BY

A. PRESCOTT FOLWELL,

Member American Society of Civil Engineers.

SECOND EDITION, REVISED.

FIRST THOUSAND.

NEW YORK:
JOHN WILEY & SONS.
LONDON: CHAPMAN & HALL, LIMITED.
1899.

Copyright, 1898,
BY
A. PRESCOTT FOLWELL.

PREFACE.

For a number of years the author has been looking for the appearance of a work on Sewerage which should embody the most recent data and ideas relating to the subject and treat of both the Combined and Separate Systems in a comprehensive manner, recognizing the fact that such a work is needed by city engineers and engineering schools. None such has appeared, and he has consequently undertaken the task of supplying the deficiency.

No attempt has been made to treat at length the subject of Sewage Disposal, for the reasons stated in Chapter II. Parts II and III on the Construction and Maintenance of Sewers will, he believes, be appreciated by those who are called upon to superintend such work without previous experience, and even, he hopes, give valuable hints to many who are not novices; although he recognizes that the ground is by no means completely covered. For much of the matter therein contained he is indebted to the engineering periodicals, particularly the *News* and *Record*, but the greater part of it has never, to his knowledge, appeared in print.

While primarily intended for practising engineers, the work has also been arranged with the idea that it may be useful as a text-book in engineering schools; Part I having already been so used by the author, and Part II having been largely given in the form of lectures to his classes.

CONTENTS.

PART I. DESIGNING.

CHAPTER I. SYSTEM TO BE EMPLOYED.

ART.		PAGE
1.	Requirements of a System	1
2.	Dry Sewage Methods	2
3.	Dry Sewage Systems	4
4.	Pneumatic Systems	7
5.	Water-carriage Systems	7
6.	Combined and Separate Systems	9
7.	Summary	12

CHAPTER II. SEWAGE DISPOSAL.

8.	"Disposal" and "Sewage" Defined	14
9.	Aims of Disposal	16
10.	Principles Involved	18
11.	Discharging into Rivers and Tidal Waters	20
12.	Sewage Treatment	26

CHAPTER III. AMOUNT OF SEWAGE.

13.	Sewerage Conduits	30
14.	Amount of House-sewage	31
15.	Data of House-sewage Flow	37
16.	Amount of Storm-water	44
17.	Rates of Rainfall	44
18.	Run-off Data	47
19.	Formulas for Storm-water Run-off	52
20.	Expediency of Providing for Excessive Storms	56

Chapter IV. Flow in Sewers.

ART.		PAGE
21.	Fundamental Theories	60
22.	Limits of Velocity	73
23.	Size of the Sewer	78
24.	Shape of the Sewer	81

Chapter V. Flushing and Ventilation.

25.	Necessity for Flushing	85
26.	Methods of Flushing	88
27.	Appliances for Flushing	93
28.	Necessity for Ventilation	95
29.	Methods of Ventilation	97

Chapter VI. Collecting the Data.

30.	Data Required	103
31.	Surveying and Plotting	106

Chapter VII. The Design.

32.	General Principles	111
33.	Subdivision into Districts	116
34.	Locating the Sewer Lines	117
35.	Volume of House-sewage	120
36.	Volume of Storm-sewage	122
37.	Grade, Size, and Depth of Sewers	131
38.	Inverted Siphons	138
39.	Sub-drains	139
40.	House and Inlet Connections	141
41.	Manholes, Inlets, Flush-tanks, etc.	144
42.	Pumping of Sewage	148
43.	Intercepting-sewers and Overflows	153
44.	Use of Old Sewers	155

Chapter VIII. Detail Plans.

45.	The Sewer-barrel	157
46.	Pipe Sewers	166
47.	Manholes, Lamp-holes, Flush-tanks, etc.	171
48.	Interceptors and Overflows	183
49.	Inverted Siphons, Sub-drains, Foundations	185

CHAPTER IX. SPECIFICATIONS, CONTRACT, ESTIMATE
OF COST.

ART.		PAGE
50.	Definition and Classification of Specifications	190
51.	Specifications for Materials	192
52.	" " Excavation	198
53.	" " Construction	202
54.	" " Back-filling and Cleaning Up	212
55.	General Provisions, Payments, etc.	216
56.	Contract	223
57.	Estimate of Cost	226
58.	Methods of Assessment	231

PART II. CONSTRUCTION.

CHAPTER X. PREPARING FOR CONSTRUCTION.

59.	Contract Work or Day Labor	237
60.	Obtaining Bids	239
61.	Engineering Work Preliminary to Construction	241
62.	Other Preliminaries	242

CHAPTER XI. LAYING OUT THE WORK.

63.	Lining Out Trenches	244
64.	Giving Grade	245

CHAPTER XII. OVERSIGHT AND MEASUREMENT OF WORK.

65.	Inspection of Work	252
66.	Duties of the Engineer	254
67.	Measurements	256
68.	Final Inspection	260

CHAPTER XIII. PRACTICAL SEWER CONSTRUCTION.

69.	Organizing the Force	265
70.	Trenching by Hand	270
71.	Excavating Machinery	275
72.	Sheathing	279
73.	Laying Sewer Pipe	291
74.	Building Masonry Sewers	297
75.	Building Manholes and Other Appurtenances	306

ART.		PAGE
76.	Foundations	309
77.	Pumping and Draining	310
78.	Handling Wet and Quicksand Trenches	315
79.	River Crossings and Outlets	328
80.	Crossing Railroads and Canals	335

PART III. MAINTENANCE.

CHAPTER XIV. HOUSE CONNECTIONS AND DRAINAGE.

81.	Necessity for Intelligent Maintenance	340
82.	Requirements of Sanitary House-drainage	341

CHAPTER XV. SEWER MAINTENANCE.

83.	Requirements of Proper Maintenance	347
84.	Flushing	349
85.	Cleaning	354

TABLES.

1.	Population and Per Capita Water Consumption in Different Cities	32
2.	Population, Number per Family and per Dwelling, Different Cities	35
3.	Gaugings of Sewage Flow, Providence, R. I.	38
4.	" " " " Toronto, Canada	39
5.	" " " " Schenectady, N. Y.	39
6.	" " " " Atlantic City, N. J.	39
7.	" " " " Weston, W. Va., Insane Hospital	40
8.	" " " and Water Consumption, Des Moines, Ia.	43
9.	Maximum Rates of Rainfall in Various Sections	46
10.	Relative Cost and Capacity of Sewers, Washington, D. C.	58
11.	Velocity and Discharge in Circular Sewers, 4 to 36 in. Diameter	64
12.	" " " " " " 33 in. to 10 ft. Diameter	66
13.	p, a, R, Velocity and Discharge for Different Depths of Sewage, Circular Sewers	69
14.	p, a, R, Velocity and Discharge for Different Depths of Sewage Egg-shaped Sewers	70
15.	Materials Moved by Different Velocities of Water	74
16.	Calculation of Sewer Sizes for Minimum Grades	134
17.	Prices and Weights, Vitrified Clay Sewer-pipe	227

ART.		PAGE
18.	Prices of Drain-tile	228
19.	" " Light-weight Iron Pipe	228
20.	Amount of Cement for Laying Different Sizes of Sewer-pipe	228
21.	Cost of Excavating, Back-filling, and Sheathing Trenches	229
22.	Cost of Laying Sewer-pipe	229
23.	Cost of Circular Brick Sewers	230
24.	Cost of Manholes	230

ILLUSTRATIONS.

PLATE		
I.	Des Moines Sewer Gaugings and Water Consumption	41
II.	New Orleans Run-off Curve Diagrams	49
III.	Plan of a House-sewerage System	123
IV.	Rainfall Diagrams and Acreage Curve	125
V.	Plan of a Storm Sewer System	128
VI.	Sections of Masonry Sewers	160
VII.	" " " "	162
VIII.	" " " and Pipe Sewers	164
IX.	Manholes and Lamp-holes	173
X.	Manholes, Flush-tanks, and Inlets	175
XI.	Interceptors, Siphons, Sub-drains, etc.	184
XII.	Trestle Excavating-machine at Work	277

FIGURE		
1.	Modified Birmingham Pail	5
2.	Egg-shaped Sewer	82
3.	Sounding-rod	109
4.	Alignment of Sewer Junctions	119
5.	Method of Setting Grade-plank	246
6.	" " " " "	247
7.	" " Holding Grade-cord	248
8.	Grade-rod	248
9.	Inspector's Templet, Egg-shaped Sewer	262
10.	Invert-former	263
11.	Excavation-platform	271
12.	Cross-staging in Trench	272
13.	Skeleton Sheathing	281
14.	Sheathing under Braces	283
15.	Driving-cap and Maul	285
16.	Horizontal Sheathing	285
17.	Sliding Rod for Measuring Braces	287
18.	Sheathing Puller	289
19.	Pipe-laying Hook	292
20.	Appliance for "Entering" Heavy Pipe	292

FIGURE		PAGE
21.	Pipe-cleaning Disk	295
22.	Templet for Brick Sewers	298
23.	Hod for Lowering Brick	300
24.	Mason's Platform for Brick Sewers	301
25.	Centre for Brick Sewers	302
26.	Form for Concrete Arch	305
27.	Sewer-pipe Laid in Concrete	319
28.	Sheathing a Badly Caved Trench	321
29.	Appliance for Cleaning Sub-drains	326
30.	Sewer Crossing Creek above Water	329
31.	Coffer-dam Puddle Walls	333
32.	Sheathing on Steep Slopes	337
33.	Flange for Pipe in Embankment	338
34.	Appliance for Cleaning Siphon-sump	355
35.	Disk for Cleaning Sewers	357
36.	Method of Using Cleaning-disk	358

SEWERAGE.

CHAPTER I.

THE SYSTEM.

Art. 1. Requirements of a System.

A SYSTEM for the removal of sewage is demanded by a populous community on two grounds: the higher one of the public health, and the more popular one of convenience; and in designing a system each of these purposes must be kept constantly in mind, the first being ever given predominance over the second if they conflict in any way. The proper meeting of these demands determines the principles of designing.

There are two imperative essentials to sanitary sewerage:

I. That the sewage, and all the sewage, be removed without any delay to a point where it may be properly disposed of.

II. That it be so disposed of as to lose permanently its power for evil.

Convenience requires that the sewage be collected and disposed of with the least trouble to the householder and in the least obtrusive and offensive way.

In taking up the study of sewerage for any particular place or community the first question arising is the general system to be adopted. In many cases financial limitations will be forced upon the engineer as an unfortunate but imperative argument in the choice not only of the details of the system but even of the system itself. He must perforce recognize these limitations in addition to the requirements of sanitation and convenience, but should not carelessly assume that since there is but little money to spend upon the work the care given to the design will need to be only proportionately great. He should realize that the highest talent is needed to obtain the best results with limited resources.

The solution of the difficulty when a complete water-carriage system is rendered out of the question by reason of its cost may lie in the construction of only the most necessary portion of the system or in the adoption of one of the dry-sewage systems.

ART. 2. DRY SEWAGE METHODS.

The methods in common use for removing excrement and liquid wastes may be conveniently divided into three general classes: (1) Dry Sewage, (2) Pneumatic, and (3) Water-carriage systems.

The most primitive method of application of excrements to the soil—if it can be called a method—would be embraced under the first head. The old-fashioned privy was a step forward, and in a large part of this country is as yet the only one which has been taken, privacy being the main argument for its adoption. But, while contributing somewhat to this and to comfort, it cannot be considered as a sanitary appliance. "Constructed for the avowed purpose of retaining the solid matters as long as possible upon the premises, they become centres of pollution and infection. The liquid portions, escaping, pollute the soil and neighboring wells; the noxious exhalations arising from their putrefying contents contaminate the air." (Samuel M. Gray's Report on Proposed Sewerage System for Providence, R. I.)

Regular movement of the bowels is essential to health and to bodily and mental vigor. Yet a rainy day, a deep snow, or publicity of location has kept many a person from the daily attention to nature's demands when this requires a visit to the outdoor privy.

This last objection is met by the indoor closet connected with a cesspool; but there is probably no subject upon which sanitarians are more thoroughly agreed than upon the inherent vileness and danger of the cesspool as ordinarily constructed. Fresh sewage if not taken into the stomach is neither injurious to health nor very offensive to the smell; but from putrescent excreta and kitchen slops come those noisome gases which, if not themselves bearers of malefic germs, at least lower the vitality and render the body more vulnerable to disease. Retained for weeks and months in a liquid or semi-liquid state in a cesspool, sewage is then under the conditions best adapted to putrefaction in its foulest form. And in very few, if any, cases is the plumbing of the house adapted to exclude from the air of the dwelling the gases emitted; indeed it is doubtful if this can be accomplished with certainty when, as is too often the case, the cesspool is tightly covered or sealed with snow or ice. Moreover, practically no cesspools are water-tight, though many are thought to be so. A cesspool $8\frac{1}{2}$ feet in diameter and 10 feet deep to which a family of five contribute a daily average of 25 gallons of sewage (a low estimate) would, if tight, require to be cleaned twice each year. Very few, it is believed, are cleaned this often; many are never cleaned, but the contained liquid leaches out into and through the adjacent soil, which soon loses its power to purify it.

This vilest of liquids is dangerous in two ways: it may reach and taint wells for hundreds of feet around, and it may pollute the air existing in the soil under cellars, which air will exhale and permeate the houses above. In excavating for sewers in gravelly soil in a city street the author has found the gravel colored black by the liquid from a cesspool located 75 feet distant in the rear of the house opposite; which liquid must consequently have passed under or around the cellar of this house.

It seems advisable to speak thus at length on this subject for the reason that many intelligent persons look with favor on the cesspool as a sanitary contrivance, whereas in most cases it is one of the greatest abominations permitted in any civilized community.

Art. 3. Dry Sewage Systems.

The methods already referred to can hardly be called systems, but are rather makeshifts. The simplest *systems* which can be at all commended are the Pail system and the Earth-closet. These are used but little in this country, but would be for many small villages a vast improvement over the privy or cesspool.

The Pail system consists essentially of the placing under the privy-seats of pails, which are to be removed, emptied in some spot where a nuisance will not thereby be created, cleaned, and returned. Duplicate pails must be provided to be used in place of these during their absence.

This method is in use at Marseilles, Havre, and other French cities; at Rochdale, Birmingham, Manchester, and other places in England; but only in certain districts of these cities, which are introducing water carriage and are yearly increasing the territory thus sewered. It has been used by a few communities in this country also, among them Vineland, N. J.; but this place also will probably replace it with a water-carriage system.

A modification of and improvement upon the Pail system is the Earth-closet system, in which pulverized dry earth, charcoal, or ashes are used as a deodorizer and are applied to the excreta while fresh, the mixture being subsequently removed, preferably as in the Pail system. Brick-clay and loam rank high as deodorizers when applied in a perfectly dry and powdered state. Ashes are not so effective. In Bremen powdered turf is used. There is not evident a sufficient superiority in charcoal to compensate for its cost and other disadvantages.

The deodorizing-powder should be applied each time the closet is used. An excellent arrangement is that of a large box or barrel resting upon an extension of the seat and with an aperture and slide so contrived that any desired amount of the powder may be deposited upon the excrement by a

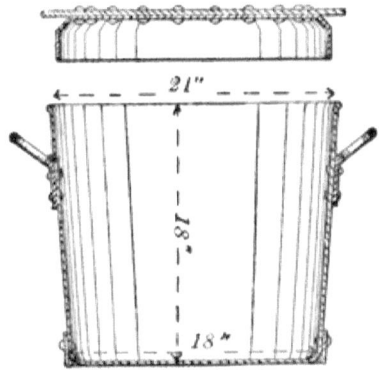

FIG. 1.—MODIFIED BIRMINGHAM PAIL.

slight motion of a convenient handle. The simplest method of applying the deodorizer is by a small scoop or shovel, the earth being kept in a box placed in a convenient position in the closet.

For either the Pail or Earth-closet system the receptacle should be round, as this form is more easily cleaned than a square one; and preferably of metal, as a wooden pail soon becomes saturated with foul liquors. A good form is that of the modified Birmingham pail. The pails should be thoroughly cleaned after each emptying. If the earth closet is used a thin layer of earth should be spread over the bottom of the pail when it is replaced under the seat.

The mixture of earth and excreta may be dried and used again; but there is a possible danger in this, since bacteria are not often destroyed by moderate heat; it will probably be found more convenient also to deposit it immediately upon the garden or field as a fertilizer. If the Pail or Dry-earth system is adopted for a village or city an arrangement may be made by contract for removing the buckets or tubs at intervals of not more than a week, the material to be disposed of by the contractor. Such disposition of it should be made —either by placing it directly upon the fields; or by drying and pulverizing it, in which form (*poudrette*) it is more convenient for use as a fertilizer; or by burning it (see Chapter II)—as will avoid the creating of a nuisance (see Art. 10).

There are several methods, some patented, for disposing of dry sewage and garbage on the premises by means of heat, by either drying or cremating. The heat for these is obtained either from a furnace constantly burning, in which case its use in summer is exceedingly inconvenient and is usually dispensed with; or by occasional fires lighted at long intervals, during which the waste matter undergoes dangerous putrefaction. On account of these and other equally serious objections these methods are not to be commended, particularly since the cost, were every house to adopt them, would in most locations suffice to construct an excellent water-carriage system.

These dry-sewage systems, though improvements on the privy and cesspool, are imperfect from a sanitary point of view in that they require the excreta to be stored about the premises for a certain period, and because they fail to provide for the removal of slops and sink-water and dispose of urine to a limited extent only. Neither do they provide for the drainage of the soil nor for the removal of surface-water. Convenience also is not fully served by their use.

Art. 4. Pneumatic Systems.

In the Pneumatic systems the fæces only are removed, the house drainage, surface- and subsoil-water requiring a separate system of sewers or utilizing the gutters. The most widely known of these are the Liernur and the Berlier—the first used principally in Holland, the second in Paris. These two are practicable under certain conditions only and will not be described at length. Their object is to remove the sewage at frequent intervals through pipes, by means of compressed air or a vacuum, to a central station, there to be disposed of in some way, usually by being manufactured into a fertilizer. The great cost of these is prohibitive to their introduction into small cities and towns, and on account of their limited applicability, as well as for practical and sanitary reasons, their adoption in future designs is improbable.

The Shone system, which is used to some extent in England and her colonies and in this country, although classed among the Pneumatic systems, is really not in itself a system, but an application to the water-carriage system of a method of pumping sewage by the direct action of compressed air. It will therefore be considered under the head of the Water-carriage System.

Art. 5. Water-carriage System.

The Water-carriage system has now been so almost universally adopted where any improvement upon the primitive privy has been attempted that the term " Sewerage System " is ordinarily used without further qualification to refer to it. When properly constructed and managed it is certainly deserving of its popularity, being the best and cheapest method yet contrived for the removal of sewage.

As its name implies, its distinctive characteristic is the removal through conduits, by gravitation, of sewage which has been greatly diluted with water. It meets the first principal requirement of a sanitary system (Art. 1)—it removes all house-wastes and removes them immediately. It also serves the secondary but by no means unimportant purpose of removing the surface-water and draining the ground. Its convenience also is excelled by no other system. Moreover, where the territory is quite thickly populated—as in the average town—it is in the end cheaper than any other system. The two most weighty arguments against it are the large amount of water needed for its efficient working, and the pollution of streams and waste of the valuable manurial properties in the sewage when this is emptied into river or sea, as is frequently done. Victor Hugo in his "Les Misérables" devotes a long chapter to the "Crime of the Century" involved in this waste. But whether this matter is ultimately wasted or its use by man only deferred it is not necessary to discuss. The all-convincing argument with any but the sentimentalist is that, while there may be manurial value in sewage, no commercially profitable method of utilizing it has yet been found. The best disposition to be made of it is therefore that which is least harmful, unpleasant, and expensive, and in most cases water carriage enables us to provide such disposition.

The argument that its proper working involves the use of large quantities of water is undoubtedly true. But where water-works already exist this objection has little force—less in this country than abroad, where 20 to 40 gallons per capita is considered a liberal allowance for water-consumption;

while in this country our small cities must provide two or three and the large ones five or six times this amount, which, with in many cases a small percentage additional for flushing, is usually sufficient and no difficulty is found in providing it. Some expense, however, is frequently incurred for flushing-water and to this extent is there force to the objection.

Places which are without a general water-supply or the general use of individual power-supplies are barred from the adoption of the Water-carriage system. For such the best plan is to adopt the Earth-closet system until such time as water has been introduced into most of the dwellings, when a Water-carriage system may be initiated, the Earth-closet pails being continually relegated, as the conduit system is extended, to the outskirts of the town, where the growth will probably keep a year or two ahead of the water-supply and sewer-construction.

Other objections are sometimes raised to the Water-carriage system which are either equally applicable to all systems or which are the result of prejudice. The possibility of the introduction into dwellings, through the house-connections, of sewer-air (which is not a "gas") is one of these, and is certainly a real one. But the resulting danger is not so great as that connected with similar evils of other systems, and it is preventable by careful designing and construction of the sewers and house-plumbing.

ART. 6. COMBINED AND SEPARATE SYSTEMS.

It is generally conceded by sanitarians that where the conditions render it possible the Water-carriage system should be adopted. This system, however, has been subdivided into the Combined and the Separate systems. The terms "Combined" and "Separate" refer to the two classes of waters which it is desirable to remove—rain-water and house-sewage. In the former system these are carried in a common conduit; in the latter the house-sewage is removed through small sewers, the storm-waters through other large ones or in the gutters, or partly in one and partly in the other.

The comparative merits of these is a theme much discussed and upon which unanimity of opinion has not yet been entirely reached. It will most probably be reached by mutual concession, for there are undoubtedly substantial arguments in favor of each. In some cases the one, in some the other, is most applicable. In many, if not a majority of, instances a judicious combination of the two will work to better advantage than either alone.

There is neither space nor necessity to quote in this work all the arguments advanced for and against each of the systems, or even to attempt to specify them all in detail, since the systems will be treated as coöperative rather than as rivals; for such is the relative position now assigned them by the best authorities. Their respective advantages under varying conditions will be treated of in Chapters III and VII. It may be well, however, to give in this connection a short statement of the points at issue between the two systems.

Either system must, if providing for storm-water, include sewers of large size—2, 3, even 15 or 20 feet in diameter. Yet during nine tenths of the time the amount of sewage flowing is no more than could be carried by pipes of from 4 inches to 2 feet in diameter.

(1) In the Separate system pipes of that size are used for the daily sewage, and thus greater velocity secured with a given amount of sewage and a given grade; consequently cleaner sewers.

(2) But even these will at times stop up, and then there may be some difficulty in removing obstructions from the pipes. Obstructions in the large sewers on the other hand can be readily reached and removed.

THE SYSTEM.

(1) In the Separate system the storm-sewers must be as large as those of the Combined and an additional small sewer be provided at a cost which increases by its full amount the cost of the Separate over that of the Combined system.

(2) But in the Separate system the storm-sewers may frequently be placed only 3 to 5 feet below the surface, while in the Combined they must be low enough to receive the house-sewage — usually from 8 to 12 feet below the surface. The resulting increase in cost would often more than cover that of an additional system of small house-sewers. Moreover, towns too poor to put in large Combined sewers can for one third to one fifth of their cost remove their daily sewage alone by means of small pipes.

(1) The Separate system usually requires large quantities of water for its perfect operation.

(2) But flushing would improve Combined Sewers also, and would probably be employed if the amount of water necessary to keep them clean were not, through its vastness, prohibitive, unless it can be introduced from a river or other large body of water, a plan which is sometimes adopted.

(1) For small sewers steeper grades are necessary than for large ones.

(2) *If each is running full;* but with a given amount of sewage the larger sewer must have the steeper grade if the same velocity is to be obtained.

(1) There is a tendency in using the Separate system to allow storm-water to run for long distances upon the street-surface.

(2) But this fault (when it is a fault) is one of the designer, and not of the system.

(2) But foul air is more diluted in large ones.

(1) Ventilation is more rapid in small sewers.

(2) But on the other hand street-washings are often as foul, though not usually as dangerous, as house-sewage, and should be purified.

(1) A very important argument in favor of the Separate system, and one which has the backing of the law in many States, is the practical necessity for its use where treatment of the house-sewage is either immediately necessary or may in future become so.

Many other arguments have been advanced on both sides, but the most weighty in favor of the Combined system are: its economy in first cost over two Separate systems, and the ease with which obstructions can be removed and a general examination of its contents made; in favor of the Separate system: its being self-cleansing, its adaptability, as a house-sewage system only, to small and poor towns, and its necessity to an economical sanitary treatment of the house-sewage.

Art. 7. Summary.

The proper conclusion in reference to the system to be adopted would seem to be—the water-carriage, where its expense is not prohibitive and the dwellings are abundantly supplied with water. In a few exceptional cases a Pneumatic system might be preferable. But better than the cesspool or privy, if the cost or the water-supply is peremptorily limited, would be a dry-sewage system—preferably the dry-earth. The last is described to a sufficient length in this chapter, as the proper conduct of it requires little else than cleanliness and faithful attention. The disposal of sewage thus collected will, however, be referred to in Chapter II.

The water-carriage system is more complicated in design, in construction, and in operation; and to the consideration of this system the remainder of this work will be devoted.

CHAPTER II.

SEWAGE DISPOSAL.

It is not intended to give a complete treatise on the subject of sewage disposal. To do so would double the size of this book; moreover, the subject has been most ably treated by others (see "Sewage Disposal in the United States," by Rafter and Baker; "Sewage Treatment, Utilization, and Purification," by J. W. Slater; "Sewage Disposal," by Wynkoop Kiersted; "Sewage Disposal Works," by W. Santo Crimp; also Reports of the Massachusetts State Board of Health). It is thought advisable, however, to give a short statement of the present knowledge and practice on the subject, since the design of the sewerage system will depend to a certain extent on the method of disposal adopted.

ART. 8. "Disposal" and "Sewage" Defined.

The word *disposal* is often used where *treatment* would be more properly employed. As a matter of fact all sewage, dry or water-carried, must be disposed of in some way after having been collected by a sewerage system. But if this disposal consists of anything other than throwing away the sewage this may be properly called a treatment thereof. These words will be thus used in this work—*disposal* as a general term, *treatment* as a more specific one.

For a proper consideration of the various methods of disposal it will be necessary to understand the results aimed at and the principles involved. And first we must understand what is implied by the word sewage. In the dry sewage and pneumatic systems it means human excreta and nothing else. In the water-carriage system, however, sewage may be found to contain almost every description of waste matter: fæces, house-" slops," manufacturing waste-waters and acids, drainage of stables, piggeries, and slaughter-houses, waste paper and rags, and frequently " swill," and numberless matters which should never reach the sewer. This is ordinarily called house-sewage. Into combined and storm sewers, besides rain-water, not only horse-droppings and vegetable refuse but sand, clay, gravel, and other heavy matters find admission through the street-inlets. These go to make what is called storm-sewage. The common impression is that of these the human excrements alone are dangerous; and this is to a large extent true so far as concerns dissemination of the germs of disease. But it is known that, aside from this, kitchen-wastes are fully as dangerous, since they contain practically the same putrescible matter, and in a state less easily rendered innocuous by either natural or artificial means. Where storm-water is admitted to the sewers the large quantities of horse-droppings which are washed in during the first few minutes of each rainstorm render the water nearly as offensive, if not so dangerous, as do human excreta.

Owing to diversity of manufacturing industries, to differences in the characters of the water used by different towns, and to other local peculiarities the sewage of each town varies from that of almost every other. Therefore the question of the proper disposal of this compound is seen to be a problem of no easy solution. The difficulty of treatment is increased by the exceeding dilution of the sewage, since the sewage of an average American town will contain but about 1 part in

1000 of organic matter, 1 part of mineral matter, and 998 parts of water.

In this chapter house-sewage only is considered, as there are very few cases where storm-sewage is purified. It is desirable, however, to connect with the house sewers cab-stands, market-places, and other parts of streets liable to collect considerable filth, small inlets being used, so that only a small amount of water from any storm can enter them, or else special traps, ordinarily closed, but through which the filth can be washed by hose.

Art. 9. Aims of Disposal.

The first aim is the getting rid of the sewage; the disposing of it in such a way and such a place that it will not create a nuisance. Communities, being even more selfish than individuals, seldom regard the well-being of other communities, but are satisfied if no nuisance is created within their own limits. It is here that the State, by its laws and through its Board of Health, should interfere for the protection of each community against all others. In England this protection is afforded by national laws and a national board. In this country many States afford a certain amount of such protection, varying from that given by the excellent laws of Massachusetts down to the almost total lack of any such protection which exists in many of even the older States. It is a duty which the engineer owes to humanity to educate the people to the importance of this matter; though he will often be compelled to yield, in part at least, to the selfish demands of those for whom he acts, that they be put to no expense for protection of other communities not required by State or national laws.

Where this protection is afforded through adequate laws properly enforced the disposal of the sewage must be such that it will "lose permanently its power for evil." How this can best and most economically be done is the question to be solved.

Many attempts have been made at a solution of this question of disposal which shall not only meet the sanitary requirements, but which shall also be financially remunerative. Some reports of success have been heard of, but when investigated the details are found to be disappointing. An English company which used a method of Chemical Precipitation was reported as paying dividends, but inquiry showed that these were but a part of the sum paid to the company by the district for disposing of its sewage, and the taxpayers were but little benefited in pocket by the method employed. Investigations of other cases have resulted somewhat similarly. The author knows of no case where the disposal of sewage is accomplished at a profit to the city or town. In the case of water-carried sewage this is not to be wondered at, since the value of the manure contained in one ton of Boston's sewage, for instance, is estimated to be but one cent.

An exception must be made, however, in the case of the Liernur system. It is reported of the manufacture of *poudrette* from a portion of St. Petersburg's sewage (collected by the Liernur system) that "it is a groundless assertion that the manufacture of *poudrette* does not cover the costs." It is possible that the force of this statement should be modified by accenting the word "manufacture." But the official reports of the city of Amsterdam (where the Liernur system is used) state that "the value of the dust-manure made from the sewage covers the whole working expense of the system and leaves a considerable margin besides." (Report of Charles Jonas, United States consul-general in 1894.) This is undoubtedly accounted for by the fact that fæces only are collected unmixed with water or unmanurial matter. As stated before, this system does not fully meet the sanitary requirements and is not adapted to this country, accustomed as we are to the abundant use of water and to modern conveniences.

The sewage of several of our Western cities situated in the "desert" region is disposed of for irrigation at a considerable profit. Los Angeles, Cal., received in 1895 a net revenue therefrom, above all salaries and repairs, of $1140, and in 1896 of $943.30. This, however, was due to the value of the water rather than of the sewage, and such an income would not be received were not water of any kind of considerable value. At Altoona, Pa., where the sewage is turned upon a farm of 56.8 acres area, the rental obtained in 1898 for the farm, including the sewage, was only $100. Few, if any, farms in a district where irrigation is not necessary, and on which the sewage must be turned in rainy as well as in dry weather, will bring any considerable rental.

It is possible that some method may be found which shall accomplish both sanitary and financial success in the treatment of sewage. But the existence of such an one has not yet been demonstrated. This being the case, the endeavor should be to find for each place that method of disposal which, under the existing conditions of location, character of sewage, etc., will best meet the requirements both of the State laws and of the laws of sanitary science, and which will be least expensive, both first cost and maintenance being considered.

Art. 10. Principles Involved.

For an exposition of the principles involved we must call upon chemistry, biology, bacteriology, medicine, and kindred sciences. Their teachings, stated generally, are:

That matter in a state of putrescence is harmful to human life if taken into the system.

That volatile emanations from such matter when breathed

into the lungs lower the tone of the constitution and render it more susceptible to, if they do not indeed directly occasion, disease.

That many diseases may be contracted by taking into the stomach certain germs which are found to be excreted by those already sick of such a disease, and these germs will exist for days in sewage having any amount of dilution.

That ordinarily sewage does not putresce until from twenty-four to sixty hours after its discharge, or even longer under certain circumstances, such as absence of moisture.

That the only true destruction of the dangerous characteristics of sewage is that effected by oxidation and by removal of the disease-germs.

That oxidation does not destroy but merely transforms the putrescible organic matter into harmless mineral compounds.

The legal principles involved vary in different localities and with different interpreters of the law, frequently depending upon the ruling as to what creates a nuisance. "I should include under this head any matter, whether solid, liquid, or gaseous, which is itself injurious to health or which may become so in contact with other substances, whether the latter may be in themselves hurtful or not; further, any matter which, though not demonstrably poisonous, is offensive to the senses." (Slater, "Sewage Treatment, Purification, and Utilization.") Such disposition of any matter that it may, while in the condition above described, approach within effective distance of any dwelling or occupied land should be held to be a nuisance. A recent ruling in the United States has included in the "creating of a nuisance" the rendering unfit for drinking purposes water which would otherwise be used thus. It is advisable both to consult the State Board of Health and to obtain reliable legal advice before deciding finally the question of disposal.

With these principles in mind a thorough and intelligent study of the local conditions should be made to decide how the requirements of sanitation and of law may best be met; whether any treatment of the sewage will be necessary, and if so which is best adapted to the given conditions.

ART. 11. DISCHARGING INTO RIVERS AND TIDAL WATERS.

The simplest solution of the problem, where it is permissible, and the one most frequently employed in this country, is to discharge the sewage directly into some flowing stream or large body of fresh water, the ocean or one of its estuaries. This is called "disposal by dilution." So far as cheapness is concerned this stands easily first among the methods of disposal, since it requires the purchase of no land and needs no care to regulate its working, excepting where the discharge is into tidal waters, when some expense is frequently gone to, both of first cost and of maintenance, to regulate the time of discharge. It is usually efficient also in removing the sewage beyond the limits of the area contributing to its volume. Looked at in a less selfish way, and considering the good of the State and country as well as of the locality sewered, other and adverse arguments present themselves. Although the sewage is removed to a distance from the contributing territory by tides or currents, it may be deposited in proximity to other communities, on banks or shores or retained by dams, thus creating a nuisance; or may render unfit for drinking, household, or manufacturing purposes water which would otherwise be so used. Under properly prepared State laws such interference with the health and rights of others should be preventable by injunction, or in the case of injury to manufacturing interests should subject the city to forfeiture of damages. An interesting case under the latter head is that decided in 1898 against

(Greater) New York City and in favor of an oysterman whose beds were destroyed by the discharge from a near-by sewer-outlet and who was awarded their value in damages.

In this connection reference should be made to the danger of spreading certain diseases through the agency of oysters, and that of the destruction of fish by sewage disposed of by dilution. There seems to be little doubt that typhoid and probably other fevers have been so conveyed by osyters, as at Brightlingsea, England, oysters from which place were accused on good evidence of having caused twenty-six cases of typhoid fever in 1897. These were exposed, however, to contact for hours at a time, at low tide, with sewage but little diluted. In view of this and of similar cases both in this country and abroad it would seem advisable that precautions be taken by the authorities to protect oyster-beds from sewage or prevent the gathering of oysters from sewage-contaminated waters.

It is probable that germs of enteric diseases are conveyed on the outside rather than the inside of the body of the oyster, and that there is little danger in eating sewage-fed fish or cooked shellfish. The health of fish does not necessarily suffer from sewage pollution of the waters they occupy, but on the contrary a moderate pollution by organic matter seems to favor their growth. (Acids, dye-house wastes, etc., are fatal to them, however.) Within a few feet of an outlet in Boston harbor from which about 40,000,000 gallons of sewage daily has been flowing for three years crabs and bivalves have been and are apparently thriving. There seems no reason to fear that organic sewage, if well diluted, will exterminate fish life from a river; in fact it is much less harmful to this than the effluent from many a sewage disposal works employing chemicals.

It is still a mooted question to what extent sewage is rendered innocuous by dilution in a flowing stream, and for how great a distance such a stream continues objectionably impure. Such changes as take place in the stream are due to dilution and sedimentation, a slow oxidation, and to a certain extent a consumption of the sewage matter by fish and lower forms of animal and vegetable life. Sewage has been traced in the water of a river for scores of miles below the point of its discharge; and instances are recorded where analyses of water taken at intervals of several miles showed practically no disappearance of the harmful organic matter, but only increased dilution. Also, since typhoid germs have been known to live in ice-water for twenty-five days, it is argued that the water of a river receiving sewage is dangerous for use as drinking-water for a distance below the sewer-mouth covered by the flow of the river during at least twenty-five days, or say six hundred miles.

Basing their conclusions on such arguments, some sanitarians maintain that house-sewage should never, or only in very exceptional cases, be discharged into a stream or lake. The arguments in favor of their standpoint are certainly weighty, but on the other hand the cost of treatment is considerable, and many towns could not afford a sewerage system at all if a plant for treatment also were necessary. A balance of benefits and evils, of what is desirable and what is possible, must be made for each case. "A question which we should be glad to have answered is this: To what extent must a polluted liquid be diluted in order to be safely used for domestic purposes? The answer, however, none can give. We do know this: it has been shown by actual experiment that the spores of some of the lower orders of vegetable organisms are very difficult to deprive of vitality; they may be frozen or heated to the boiling temperature, or they may be kept in a dry condition for years, and then, if placed in a favorable medium, become active and produce their kind. Admitting the presence of disease-germs in a liquid, the liquid may be diluted until the chance of taking even a single germ into the system is so small that it may be disregarded; and yet if the prevailing theory be true a single germ if taken *might* produce disastrous results. It is easy to push the demands for purity to an absurd extent; all reasonable precautions should be taken to insure purity, but there is a point beyond which it is foolish to attempt to go. In the present state of our knowledge we should, however, err on the side of safety, and the mere fact that chemical analysis fails to detect impurity should not be accepted as a guaranty that a water is fit to drink." (Nichols, "Water Supply, Chemical and Sanitary.")

"The minute forms of animal life are thus seen to be powerful agents in the self-purification of sewage-polluted water, but the conclusion which has been drawn, that therefore sewage-polluted streams are after a few miles' flow fitted through the action of such and other natural agencies for drinking, is not wholly justified by the present state of knowledge of the subject as a whole.

"Along with our knowledge of the purifying action of the minute animals and plants has grown up a more definite knowledge of the causation of typhoid fever, cholera, and the other water-borne communicable diseases; and before it can be positively affirmed that a sewage-polluted stream is safe for drinking after a few miles' flow it must be shown so definitely as to be beyond question by those whose special studies have fitted them for intelligent judgment that the purifying agencies have practically eliminated the germs of the water-borne communicable diseases. Until such showing is clearly made the proposition that crude sewage ought not to be turned into running streams, ponds, lakes, or other bodies of water which either are or may be the sources of water-supplies must be considered as holding good." (Rafter and Baker, "Sewage Disposal in the United States.")

The above restrictions apply equally to ice which may be used for drinking-water, since it is known that bacteria are not excluded from water by freezing and that many varieties will live in ice for months.

Apart from the danger of communicating disease by sewage pollution of water is that of causing a nuisance by so considerable a pollution that the combined action of dilution, oxidation, and consumption by animal and vegetable life is not sufficient to prevent a considerable sedimentation, followed by nocuous decomposition or objectionable minor evidences of the impurity of the water. Dilution is largely an expedient for facilitating oxidation, and hence the greater the amount of oxygen in a body of water the more sewage can it assimilate, both because of the greater possibilities of oxidation and because of the greater quantity of animal life present which will assist in the purification. A stream with a rapid, agitated flow may be charged with a greater percentage of sewage than a sluggish one, both because of the greater amount of oxygen and because the mixture of the sewage and water becomes more thorough. Sewage discharged into a slow current can often be traced for a long distance as a separate stream, mingling but slowly and along its edges with the purer water. A slow current will also favor deposits of the heavier sewage matter. For these reasons the point of most rapid flow of stream or current should be chosen for the outlet.

Authorities differ as to the minimum amount of dilution necessary, but this is usually placed between 1500 and 3500 gallons per day per person contributing to the sewage. The proportion is sometimes stated in terms of cubic feet of sewage, but, since the amount of impurity is not increased by greater per capita consumption or waste of water, the former method seems preferable.

In a few States sewerage systems must be so designed,

before meeting the approval of the State Boards of Health, as to permit and provide for a treatment of the house-sewage at some future time, although temporarily they may be allowed to discharge into adjacent streams. It is probable that before very long this will be the regulation in most States. But in any event where the discharge is into a stream or lake the possibility of the necessity arising in the future for treatment of the sewage should be foreseen and provided for in the design of the system.

The discharging of sewage into tidal waters involves the principles given as applying to discharge into rivers so far as creating a nuisance is concerned, and also the practical consideration of the movements of prevailing winds and tides. That is, such a location, time, and method of discharge must be selected as will insure the sewage being carried *away* from the shore. Discharging into a landlocked bay or during a flood tide will *not* insure this. An illustration of the former error is the city of Havana, Cuba, which discharges its sewage into a harbor almost surrounded by land and where tides have but little effect. Partially, at least, from this cause it results, as one of her citizens has stated, that Havana is "the most intense focus of yellow fever in the world." "In every case the outfall of the discharging sewer should be below the level of the water at all states of the tide, and be provided with a tidal valve, to prevent the ingress of sea-water. The position of the outfall should, if possible, be so chosen that the sewage will be always carried out to sea independently of the tides and the possibility of its return avoided; and for this purpose advantage should be taken of any current that flows off or along the shore, the sewage being discharged into it, and thus carried away from the neighborhood of the town. If there is a current setting along the shore, then the sewer outfall should be placed at that extremity of the town which will prevent the sewage being borne along the whole sea front.

The prevailing winds must also be taken into account, so that floating matters may not be blown back toward the town." (W. H. Corfield, " Treatment and Utilization of Sewage.")

The desired result may usually be obtained by discharging into an ebbing tide only. In general also it may be said that the deeper the outlet of the sewer the less the probability of floating matter being blown upon the shore, because the greater its comminution.

Art. 12. Sewage Treatment.

In many inland cities no streams of sufficient size are available as outlets for the sewers, and these, as well as those which for other reasons do not adopt dilution as the method of disposal, must needs treat their sewage in some way. The methods of disposal coming under the head of treatment can be generally classified as oxidation by land and by fire and precipitation by chemicals.

The chemicals most commonly used are lime and sulphate of alumina, which are mixed with the raw sewage. The object of their use is to separate the impurities from the water and thus render them more easily dealt with. A part only, though in some cases a large part, of the impurities are thus removed. Chemical precipitation involves the use finally of either earth or fire; for it but removes and does not destroy the objectionable matter, which must still in its more condensed form be disposed of.

Chemical precipitation is sometimes employed as a preliminary to land treatment, especially where the amount of land available is limited. In many cases no further treatment is given the effluent; but if additional purification is necessary a much less area of land is required for treating the effluent than would be needed to treat the raw sewage. Chemical treatment is therefore particularly applicable to localities where sufficient land for filtration or irrigation is unattainable or prohibitively expensive.

The matter precipitated is usually compressed into "sludge-cakes," which are sold for fertilizer if possible; otherwise they are disposed of on land or burned in a cremator.

During the last few years another method of treating sewage has come into prominence under the name of the septic tank treatment. In this the structure of many organic matters is broken down and resolved into carbonic acid and other gases, ammonia and other nitrogenous compounds, in the almost total absence of oxygen, the gases being allowed to escape, while the effluent is in a condition more readily susceptible to rapid purification or dilution. Sufficient data have not yet (1898) been collected to demonstrate the value of this method nor its economy when used upon a large scale.

The treatment with earth includes all use of the sewage as a fertilizer, whether in its crude state or in the shape of *poudrette*. Excrement removed by the Pail and Dry-earth systems is usually spread as a fertilizer upon land. This is commonly done by ploughing furrows about two feet apart, depositing the matter in them, and turning the mould upon it by intermediate furrows. Two such applications, in the spring and fall, are all that most lands will receive to advantage. The sludge produced by chemical precipitation is frequently used as fertilizer, or even, when it cannot otherwise be gotten rid of, is buried in cakes a few feet under the surface of the ground.

The most common use of earth, however, is in the treatment of sewage by irrigation and by filtration. The examples of this use of land are already numbered by the hundreds, with the list being constantly swelled. In these the sewage from the water-carriage system is turned upon the land, and passes downward through it, the effluent being usually intercepted by drains and conducted to a neighboring watercourse. During its transit the sewage is purified by oxidation of its organic constituents, the agent thereof being the air contained in the interstices of the soil, assisted by certain bacteria. A very large percentage of the disease-germs present are usually removed at the same time. If vegetable growth of some kind is raised upon this land the method is called irrigation; if the land is kept clear, in which case more sewage can be treated on a given area, the process is called filtration. A porous soil is that best adapted to this treatment, though many fairly successful systems are in operation on somewhat clayey or boggy soil.

Where suitable soil is not obtainable or is expensive coke-breeze or similar coarse material may be used to advantage, not more than one tenth the area required by sand filtration being necessary to obtain equal results, judging by recent experiments. Both sand and coke-breeze filtration are *filtration* in only a limited degree, and these filters are by some called " bacteria-tanks," by others " biological filters."

Disposal by fire, or incineration, is used for the destruction of house-sewage or of cakes of sludge. Cremators similar to those used for garbage destruction are excellently well adapted to this purpose; in fact where large amounts of sewage are to be disposed of by incineration some such furnace, which will consume its own gases, is necessary (see also Art. 3).

For details concerning the various methods of treatment above mentioned the works previously referred to are recommended. If treatment is necessary the planning of the same should be made a special study; and the cases are few indeed where some provision for future, if not for immediate, treatment is not imperatively demanded. The designing engineer should hesitate long before adopting as a *final* solution of the question the method of simple dilution. In general it may be said, however, that this is temporarily permissible when the water of dilution is not used as a beverage by any community within 20 to 50 miles down stream from the sewer-outlet, and provided that a dilution at least equal to that given in Art. 11 is immediately obtained, and that the current is of proper strength and direction to prevent deposits.

CHAPTER III.

AMOUNT OF SEWAGE.

Art. 13. Sewerage Conduits.

The object of a system of sewers is in general to conduct all excreta and fouled waters from the places of their origin to an appointed outlet, and as rapidly and continuously as possible. No part of the sewage should be retained in any portion of the system for any considerable time, either in its liquid form or in the shape of deposits upon the bottoms or walls of the conduits or their appurtenances; for such retention may permit of putrescence of the organic matter before it reaches the place assigned for disposal, the conduits thus becoming no better than "elongated cesspools." The insuring of this result with the greatest certainty and economy is the prime requisite in the design of a sewerage system.

The largest part of the system is made up of conduits of various size, shape, grade, material, and depth below the ground-surface. The two last are practical points to be considered later (Articles 37 and 45), but the size, shape, and grade are to be determined—approximately at least—by theoretical considerations. The data used in these considerations are (1) the amount and character of the sewage to be removed, and (2) the relative surface elevations and grades along the line of the proposed sewer-conduit. The latter are obtained by the instrumental field-work, to be discussed in Chapter VI. While the grade of the sewer need not be that of the street-surface, it cannot depart far from this without greatly increasing the difficulty and cost of construction. The two grades will therefore be approximately parallel unless very good reasons to the contrary exist.

ART. 14. AMOUNT OF HOUSE-SEWAGE.

The obtaining of satisfactory figures for the amount of sewage is one of the most difficult tasks entering into the designing of a system. The sewage to be considered is of two entirely different kinds, from two totally different sources: house-sewage from dwellings, stores, factories, and other buildings, and storm-water from the streets, the ground-surface, and from roofs. The former is limited in quantity largely by the number of inhabitants and industrial establishments and the water contributed to the sewers by each. The latter is limited by nature's local limit of intensity of rainfall, the area tributary to the sewer, and the proportional run-off.

Considering first the house-sewage, this is almost entirely composed of water which has first been introduced artificially into the dwellings or establishments. Excreta and solids legitimately finding their way to the sewer comprise only a very small part of the sewage.—from 5 to 15 parts in 10,000. There may be besides this comparatively small amounts of leakage of ground-water, roof-water, and flushing-water reaching the sewer. It would seem, therefore, that we may make a close approximation to the amount of house-sewage by using the water-consumption of the town in question. This can usually be obtained from the pumping records, or, in the case of a gravity supply, from a meter set in the main near the reservoir. Table No. 1 shows the rates for a number of cities of the United States at intervals of 10 years. This table shows the great difference between the per capita

Table No. 1.

Cities.	1870.		1880.		1890.	
	Population.	Per Capita Consumption.	Population.	Per Capita Consumption.	Population.	Per Capita Consumption.
New York City	942,292	90.2	1,206,590	78.7	1,515,301	79
Chicago, Ill.	298,977	62.32	503,304	114.0	1,099,850	138
Philadelphia, Pa.	674,022	55.11	847,542	68.1	1,046,964	131
Brooklyn, N. Y.	396,099	47.16	566,689	54.2	806,343	72
St. Louis, Mo.	310,864	35.38	346,000	72.1	451,770	72
Boston, Mass.	250,526	60.15	416,000	92.0	448,477	80
Cincinnati, O.	216,236	40.0	256,708	75.9	296,908	112
Cleveland, O.	92,829	33.24	....	65.0	261,353	103
Buffalo, N. Y.	117,714	58.08	....	106.0	255,664	186
Detroit, Mich.	79,577	64.24	118,000	152.0	205,876	161
Louisville, Ky.	100,753	29.0	....	52.0	161,129	74
Columbus, O.	....	.....	....	.....	88,150	78
Paterson, N. J.	....	.....	....	.....	78,347	128
Fall River, Mass.	....	.....	49,430	30.1	74,398	29
Cambridge, Mass.	....	.....	....	.....	70,028	64
Troy, N. Y.	....	.....	....	.....	60,956	125
Des Moines, Ia.	....	.....	....	.....	50,093	55
Erie, Pa	....	.....	....	.....	40,634	112
Terre Haute, Ind.	....	.....	....	.....	30,217	83
Wilmington, N. C.	....	.....	....	.....	20,056	22
San José, Cal.	....	.....	....	.....	18,060	194
Keokuk, Ia.	....	.....	....	.....	14,101	78
Brookline, Mass.	....	.....	....	.....	12,103	73
Baton Rouge, La.	....	.....	....	.....	10,478	19
Nanticoke, Pa.	....	.....	....	.....	10,044	199

rates in different cities. It also shows in each city an increase of from 10% to 100% in consumption during each decade. Neither the increase of per capita consumption nor the difference in rates of increase in the various cities seems to follow any law, except that the former shows a constant advance. It might be expected that the per capita consumption would be greater in cities where there was considerable manufacturing or many well-kept lawns than where these conditions did not exist; and this is the general rule—with many exceptions, however. Also large cities usually have a higher rate than small ones; but this rule also has many exceptions.

For each particular case the daily consumption should be obtained from the water-works record, or, if there are no records of consumption for that locality, a careful selection should be made of the per capita consumption of a city whose conditions closely resemble those of the place in question. From these the per capita rate will be obtained. In order to be on the safe side the present rate should be increased by at least 25% to allow for a probable increase in consumption, since the construction must serve not only the present population, but that of the next 30 or more years.

If meters are used on a majority of the services a great reduction in the consumption can be effected—from 30% to 60% in most instances.

Unless water-meters have become generally established and accepted, however, no allowance should be made for the reduction in sewage due to their use unless the average daily rate exceeds 100 gallons per capita. There is no reason for a daily rate exceeding this amount, and the present tendency is to meter supplies before they reach this point. An allowance of 100 gallons will be made in calculations in this work, as being a safe one for any but exceptional cases.

The average daily consumption, however, " is not uniform throughout the year, but at times is greatly in excess of the average for the year and at other times falls below it. It may be 20% or 30% in excess during several consecutive weeks, 50% during several consecutive days, and not infrequently 100% in excess during several consecutive hours." (J. T. Fanning, " Water Supply Engineering.") Many waterworks engineers use 75% excess as an average. This gives for a maximum flow, on a basis of 100 gallons daily, a rate of 175 gallons per capita daily $= .1215$ gallons per minute $=$ .00027 cubic feet per second.

It must be most urgently insisted, however, that each case should be studied by itself in the light of all the data available. These figures are given as approximate averages only, to be used in designing when no local records exist. It should also be borne in mind that the consumption given is an average including that used in manufacturing and for all other purposes. These last constitute a very uncertain portion of the whole, but unless there were definite figures obtainable it would not be safe to reduce the average by more than 10% to obtain a rate for residences only. As the assumed maximum rate—175 gallons—was but a roughly estimated average, it may be used unchanged for residential districts; and where factories are to be provided for a study should be made of the processes employed in them in order that a close approximation may be made to the amount of sewage to be expected from each.

The amount of house-sewage from buildings (other than waste water from factories and water-motors) which will reach any particular sewer will depend almost wholly upon the number of persons contributing to this amount. For a district or city this number may be obtained in two ways —by estimating the ultimate number of residences and assigning a certain number of occupants to each, doing the same with factories, stores, and other buildings; or by estimating the probable ultimate population per acre for different sections of the city. The former is the more accurate for built-up sections; the latter sufficiently so for undeveloped territory or that which will probably undergo a change in the character of its buildings.

For use in calculating by the first method the following table adapted from the U. S. census of 1880 is given.

There are in each city certain districts in which the population is much more dense than is indicated by this table. One hundred persons in one dwelling is not an exceptional rate in certain portions of New York City. For an ordinary residence district six persons to each dwelling is a

Table No. 2.

Cities.	Population.	Persons in a Family.	Persons in a Dwelling.
New York City	1,206,299	4.96	16.37
New Orleans, La.	216,090	4.77	5.95
Providence, R. I.	104,857	4.52	7.41
Kansas City, Mo.	55,785	5.97	6.48
Nashville, Tenn	43,350	5.09	6.13
Denver, Colo.	35,629	5.99	6.75
Harrisburg, Pa.	30,762	4.78	5.16
Erie, Pa.	27,737	5.24	5.66
Des Moines, Ia.	22,408	5.14	5.37
Sacramento, Cal.	21,420	4.51	5.07
Springfield, Ill.	19,743	5.04	5.60

sufficient average. In factories and stores which do not use water for manufacturing purposes the maximum hourly rate per capita of occupants is not nearly as great as in the case of residences; a maximum rate of 20 gallons per day will be sufficient allowance for ordinary cases, being contributed by water-closet flushes, urinals, and wash-basins. One person to each 50 square feet of floor-space may be taken as a maximum density for factories, and one to each 75 square feet for office-buildings.

A method frequently used is that of adding a percentage of increase to the present population of each city or section. American towns under 50,000 population have been found as a general rule to double in size in about 15 or 20 years. Having ascertained for each case its past rate of increase and present population, these are taken as the basis for calculations. But this increase is far from uniform over the entire area of a town, differing in different sections; also after a section has reached a certain density of population it remains practically stationary, unless its character change—as from residential to business or manufacturing. The percentage of total growth of a town may be used, however, as a check upon the sum of the populations assumed for the various sections. The law of increase varies in different cities, but that followed in the past by the one under consideration having been obtained from the records can be projected into the future, it being assumed that this law will remain constant (see Art. 129).

Considerable judgment must be used in locating division-lines between sections and assigning to each its density of ultimate population. The most hilly sections will probably be least thickly, and those in the level bottom lands most thickly, populated. Further than this it would be unsafe to try to state any general law. The least population which should be assigned to any habitable section within city limits is 20 per acre. The per acre population in any residence section can be expressed by the equation

$$P = \frac{43560 lbo}{fd[lb + w(l + b + w)]},$$

in which $l =$ the average length of a city block;

$b =$ " " breadth " " " "
$o =$ " " number of occupants of each lot;
$f =$ " " " " front feet to a lot;
$d =$ " " depth of a lot;
$w =$ " " width of a street;
$P =$ " " population per acre.

For a section where the blocks are 400 ft. by 200 ft., streets 66 ft. wide, lots 50 ft. by 100 ft., and the population residential ($o = 6$),

$$P = \frac{43560 \times 400 \times 200 \times 6}{50 \times 100[400 \times 200 + 66(400 + 200 + 66)]} = 34 \pm.$$

For a tenement district, each building on a lot 50 ft. by 100 ft. and containing on an average 80 occupants, P would equal 453, which is about P for the Tenth Ward, New York City.

A block with lots 25 feet by 80 feet and with 6 occupants each represents fairly well the most dense residence section of an average city of 10,000 to 100,000 population. This gives $P = 85$. In many cities the maximum does not exceed 50 per acre.

The population found times .00027 for residences and times .00003 for factories and office-buildings (on the basis of the previously assumed daily consumption) will give in cubic feet per second the maximum amount of house-sewage from buildings to be expected. To this must be added manufacturing wastes, which are to be allowed for in quantities which must be decided upon separately for each individual case. Also if the soil is inclined to be wet at a depth less than that of the proposed sewer (and this includes a larger proportion of localities than most persons realize) an additional allowance must be made for ground-water leaking through the joints (see Art. 35). With care this need not amount to more than one cubic foot per second for each 30 to 100 miles of sewer; but it has been known under most unusual conditions to more than equal the entire capacity of the system.

Where flush-tanks are used (see Chapter V) an additional allowance is frequently made for water from them. But this seems entirely unnecessary, since their very purpose is to temporarily gorge the sewer for as great a distance as possible; and the smaller the sewer the better is this mission fulfilled. The average discharge per minute of 100 tanks, each discharging 300 gallons once in 24 hours, would amount to only 1½% of the capacity of a 15-inch sewer at minimum grade.

Art. 15. Data of House-sewage Flow.

Instead of using rates of water-consumption as equivalent to the sewage discharge it would undoubtedly be preferable to establish the actual relation between these, based upon the rate of flow of sewage itself in various towns already sewered. But very few such records exist—too few to enable us to deduce a definite law from them with certainty, although a study of even these few is instructive. Probably the most extended series of gaugings of sewage discharge made in America are those of the Providence, R. I., sewers by Samuel M. Gray. A condensed summary of them and of other gaugings is given in the following tables:

TABLE No. 3.

SUMMARY OF RESULTS OF WEIR MEASUREMENTS OF SEWAGE FLOW IN PROVIDENCE, R. I.

(Condensed from a Report by SAMUEL M. GRAY on the Sewerage of Providence.)

Street.	Houses Connected.	Population Connected.	Average Discharge per Second.	Maximum Discharge per Second.	Date of Measurement.
Dorrance	772	6562	7.32	11.65	{ Average of May and June
Brook	575	4480	5.92	6.78	Sat., Feb. 2
"	....	....	5.61	6.78	Mon., " 4
"	....	....	3.985	5.47	Tues., July 1
"	....	....	3.88	5.47	Thurs., " 3
"	....	....	3.86	5.76	Sat., " 5
"	....	....	4.28	5.47	Mon., " 7
Elm.........	1114	8800	4.15	7.90	" Jan. 28
"	....	....	3.69	7.90	Tues., " 29
"	....	....	3.317	6.32	Wed., June 4
"	....	....	3.37	6.32	Fri., " 6
"	....	....	3.10	5.11	Thurs., " 19
N. Main.....	....	....	2.57	4.45	Mon., May 12
"	....	....	2.46	3.80	Wed., " 7
"	....	....	1.76	3.25	Fri., July 25
Blackstone....	....	....	1.50	3.20	Mon., Feb. 11
"	....	....	2.40	5.40	Thurs., " 14
"	....	....	2.30	4.82	Sat., " 16
"	....	....	2.06	2.65	Mon., " 18
"	....	....	2.106	3.20	Fri., " 29
Ives.........	204	1814	0.753	1.02	Mon., March 3
"	....	....	0.854	1.38	Wed., " 5
"	....	....	0.695	1.30	Mon., Aug. 25
"	....	....	0.600	0.92	Wed., " 27
College.......	108	824	1.05	1.82	Fri., May 2
"	....	....	1.07	1.81	Mon., " 5
Point	321	2729	1.57	3.82	{ Mean for May 6–21
Power........	31	239	0.26	0.56	Wed., April 23
"	....	....	0.26	0.54	Fri., " 25
"	....	....	0.045	0.14	" Aug. 22
Nash	25	193	0.037	0.135	Mon., April 21
"	....	....	0.030	0.060	Tues., " 22
"	....	....	0.036	0.086	Aug. 6, 7, 11
Park.........	21	162	0.034	0.187	Fri., April 18
"	....	....	0.043	0.385	Mon., Aug. 18
Martin	153	1178	1.208	1.400	Wed., July 9
Pitman.......	86	655	1.202	2.380	Mon., April 28
"	....	....	0.744	1.380	Wed., " 30

TABLE No. 4.

GAUGINGS MADE IN TORONTO, CANADA, IN THE SPRING OF 1891, LASTING THREE DAYS.

Population per Acre.	Total Population.	Discharge. Gallons per Head per Day.	Population per Acre.	Total Population.	Discharge. Gallons per Head per Day.	Population per Acre.	Total Population.	Discharge. Gallons per Head per Day.
15.7	39,014	77	17.6	6,160	101	41.7	11,300	68
46.2	17,186	133	42.3	11,125	69	9.4	7,238	105
8.8	3,168	83	39.8	6,368	113	45.7	14,213	89
44.0	572	316*	42.4	8,268	89	38.3	19,265	53
45.5	4,595	77	11.8	8,732	102	24.0	......	87
41.8	1,045	113	43.7	9,832	89			

* No explanation given for this high average.

TABLE No. 5.

GAUGINGS MADE IN SCHENECTADY, N. Y., WEDNESDAY, FEBRUARY 5, AND THURSDAY, FEBRUARY 6, 1892—HOURLY FOR 24 HOURS.

(Fifteen miles of sewers, about 1500 house-connections tributary to the point where gaugings were made. Before house-connections were made a seepage of 60,000 gallons per day was measured. There was also 50,000 gallons of water contributed daily by flush-tanks. These two, or 110,000 gallons per day, have been deducted from the total hourly flow in obtaining the quantities in the table.)

WEDNESDAY, FEBRUARY 5, 1892.

Hour	9 A.M.	10	11	12 M.	1 P.M.	2	3	4
Total flow per hour	35,217	38,769	32,892	32,892	34,049	35,217	36,490	34,049

Hour	5 P.M.	6	7	8	9	10	11	12
Total flow per hour	32,892	31,840	31,840	31,840	29,301	29,301	29,301	28,135

THURSDAY, FEBRUARY 6.

Hour	1 A.M.	2	3	4	5	6	7	8
Total flow per hour	28,135	28,135	25,711	28,135	26,853	26,833	29,461	31,840

Average flow per hour, 31,213 gallons; minimum flow, 25,711 gallons; maximum flow, 38,769 gallons, or 24% increase over the average.

TABLE No. 6.

WATER-CONSUMPTION AND SEWAGE FLOW, ATLANTIC CITY, N. J., DECEMBER, 1891–NOVEMBER, 1892.

(Average Daily Percentage of Excess of Water Consumed over Sewage Pumped—by Months.)

December.	January.	February.	March.	April.	May.	June.	July.	August.	September.	October.	November.
32	37	53	54	61	64	36	11	36	75	66	18
50	Excess of water-taps over sewer-connections, percentage										38

The average daily percentage of excess for the year = 45%.
The average excess of water-taps for the year = 44%.

TABLE No. 7.

RESULT OF A GAUGING BY WEIR MEASUREMENT OF THE FLOW OF THE MAIN OUTFALL SEWER OF THE STATE INSANE HOSPITAL AT WESTON, W. VA., IN JANUARY, 1891.

(Made by GEO. W. RAFTER. Condensed from "Sewage Disposal in the United States." Self-closing fixtures were used in the building. 10,000 gallons per day, or 7 gallons per minute, of water of condensation from the steam-heating apparatus was discharged into the sewer.)

Day.	Hour.	Rate in Gallons per Min.	Day.	Hour.	Rate in Gallons per Min.	Day.	Hour.	Rate in Gallons per Min.	Day.	Hour.	Rate in Gallons per Min
Wednesday.	12	78.75	Thursday.	1 A.M.	44.55	Thursday.	1 P.M.	99.45	Friday.	1 A.M.	34.65
	1 P.M.	93.60		2	44.55		2	86.40		2	34.65
	2	93.60		3	44.55		3	93.60		3	39.60
	3	75.60		4	44.55		4	75.60		4	49.05
	4	75.60		5	49.05		5	93.60		5	49.05
	5	70.20		6	64.80		6	81.00		6	58.50
	6	75.60		7	75.60		7	93.00		7	70.20
	7	75.60		8	86.40		8	58.50		8	86.40
	8	70.20		9	105.30		9	53.55		9	105.30
	9	58.50		10	117.90		10	53.55		10	86.40
	10	53.55		11	93.60		11	44.55		11	86.40
	11	44.55		12	93.60		12	44.55		12	105.30
	12	44.55									

The flow of the Compton Avenue sewer, St. Louis, Mo., was gauged hourly from March 15 to 23, 1880. The minimum flow measured was 88 gallons per minute, the maximum 203 gallons, and the mean 132 gallons.

A gauging of the College Street sewer, Burlington, Vt., taken at 15-minute intervals from 7.30 A.M. to 10.30 P.M. gave two maximums, one at 7.45 A.M., the other at 9 A.M.; these were each 140 gallons per minute. The minimum flow was 65 gallons, the mean 115 gallons. Fifty-four houses were connected; the tributary population was 325.

Gaugings made at Memphis, Tenn., for one day gave a maximum of 80 gallons and a minimum of 35 gallons per minute.

Gaugings made at Kalamazoo, Mich., in 1885, from 1 A.M to 12 midnight on Monday, March 9, gave a minimum

AMOUNT OF SEWAGE.

Plate No. I.

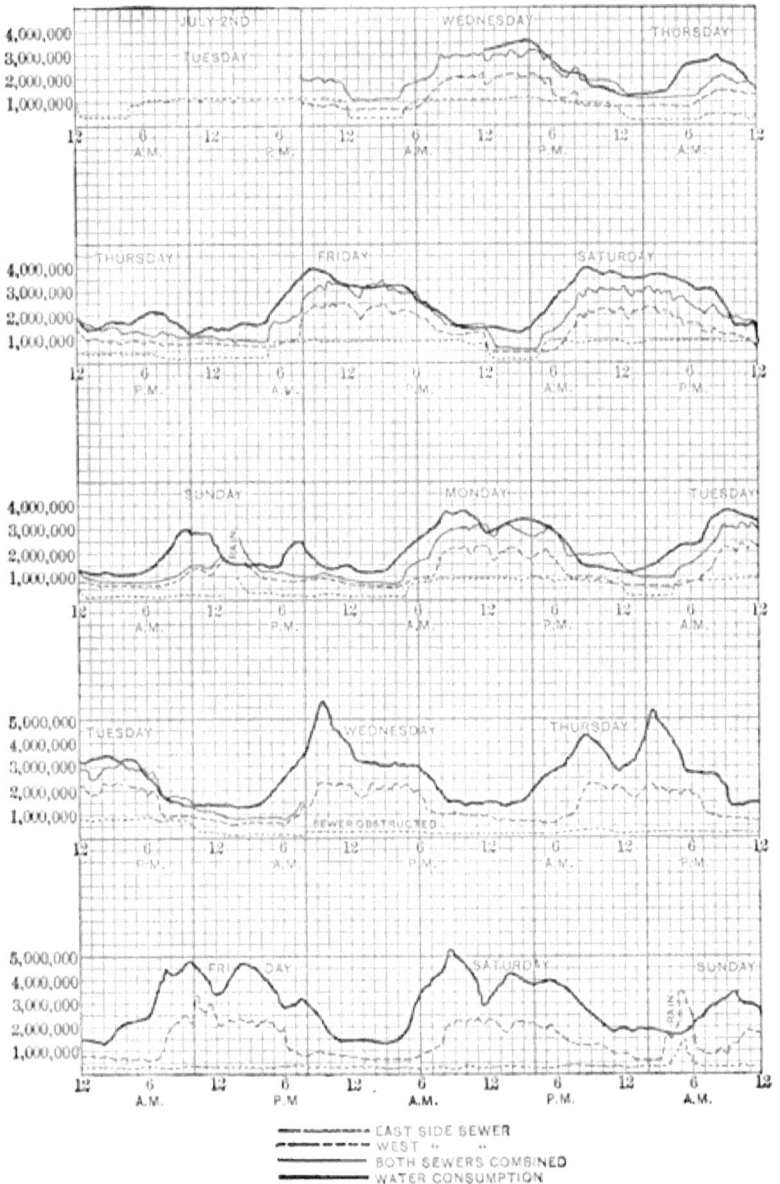

flow of 224 gallons per minute, a maximum of 287 gallons, and a mean of 254 gallons.

Gaugings were made at Des Moines, Iowa, from June 30 to July 16, 1895, by J. A. Moore and W. J. Thomas, class of '95, Iowa Agricultural College (see Plate I). The sewerage system at the outlet of which the gaugings were taken comprised: on the west side 235,000 feet of sewers, contributary population 19,400, 15 hydraulic elevators; on the east side 29,000 feet of sewers, contributary population 8100, 3 hydraulic elevators.

These were combined sewers. Rain fell on two Sundays only, and is indicated by the unusual height of the curve. Water-meters were used on the services; water was supplied to 33,700 persons, but the amount consumed by each was not ascertained, the average consumption for the city being taken. The diagram for the west side shows noon-hour stops of factories. The high-water curves for July 10, 11, 12, and 13 were caused by the water company flushing dead-ends outside of the limits of the sewers gauged. On the 12th the large flow in the west-side sewer was probably caused by a part of this flushing-water reaching it.

The maximum dry-weather rate of flow on the west side was at 10.15 A.M. Friday, July 12—175.3 gallons per capita.

The maximum dry-weather rate of flow on the east side was at 6.30 P.M. Tuesday, July 2—142 gallons per capita.

The minimum dry-weather rate of flow on the west side occurred at 4 A.M. Saturday, July 6—23.2 gallons per capita.

The minimum dry-weather rate of flow on the east side occurred at 4.30 A.M. Friday July 5—22.5 gallons per capita.

The average dry-weather rate of flow on the west side was 66 gallons per capita.

The average dry-weather rate of flow on the east side was 74 gallons per capita.

Table No. 8 gives the water pumped and the sewage dis-

Table No. 8.

Date.	Water.	Sewage.	81.6% of the Water Pumped.	Remarks.
July 3	2,720,000	2,200,000	2,219,520	
" 4	1,829,000	1,330,000	1,492,464	Holiday
" 5	2,352,635	2,050,000	1,919,750	
" 6	2,750,205	2,040,000	2,244,167	
" 7	1,809,110	1,115,000	1,476,234	Sunday; rain
" 8	2,379,820	2,030,000	1,941,933	
" 9	2,437,825	2,020,000	1,989,265	

charged during the seven days when the measurements taken were apparently reliable. The first column gives the total amount of water pumped; the second, the total sewage flow; the third, 81.6% of the first column, that being the proportion between the number of water-taps and that of the sewer-connections. The close correspondence between the two last columns shows what an excellent index the water-consumption furnishes, in this town at least, of the total house-sewage to be expected. July 12 was the only date when the maximum flow of the west side exceeded 125 gallons per capita, and then for two hours only. The average for this side was 66 gallons. Disregarding the maximum of the 12th instant, which was due to hydrant-flushing, we have a maximum for this side 89% greater than the average; and for the east side the maximum was 92% above the average. The record, however, covers two holidays out of the seven, making the average unusually low; also the general average for that time of year would ordinarily be lower than that for an entire year.

It is thought that these include all the published records of gaugings of sewage flow which have been made in this country. They seem to point uniformly to the conclusions already stated—that the winter flow of sewage is greater than the summer; that the maximum and minimum flow do not ordinarily vary from the yearly average more than 75%, but frequently do by 50%; that the house-sewage per capita very nearly equals the water-consumption where the taps and sewer-connections are equal in number.

The engineer must select and use with a great deal of judgment all the data obtainable in fixing upon the quantities which the sewer should be designed to carry. The method of making the calculations will be explained more at length in Chapter VII.

ART. 16. AMOUNT OF STORM-WATER.

The amount of storm-water reaching a given sewer depends upon the rate of rainfall, the time during which this rate is continued, the proportion of the rainfall which flows off, and the time taken by a raindrop after falling to reach the point under consideration. This last depends upon the shape, extent, and nature of the surface over which, and the length and grade of the sewer through which, it must flow.

ART. 17. RATES OF RAINFALL.

It is apparent that the rate at which the water reaches the sewer depends to a greater or less degree on the rates of rainfall from minute to minute, and not upon the amount falling in a day or even in an hour. Records giving rainfalls for these latter units of time are, therefore, valueless to the sewerage engineer. It is only within recent years that gauges have been used which automatically register the rate of rainfall at each moment of a storm. But so great a necessity for such records has been felt that the use of self-registering raingauges is becoming more and more general, and in most of the large cities continuous record of the rates of rainfall is obtained either by a city department or by the United States Weather Bureau.

Since the maximum amount of water to be removed determines the size of the sewer, we are concerned only with the maximum rates or those near the maximum. Rates of heavy rainfalls for various cities of the United States are given in Table No. 9. Where possible several high rates during the same or consecutive years are given for each locality, but no attempt has been made to give a record of all severe storms for any one place or year. An examination of rainfall data covering many years shows that in New England a rate of 3.6 inches an hour continuing for 5 minutes may be expected every year or two, a rate of 2 inches continuing for 20 minutes, a 1.5-inch rate continuing for 30 minutes, and 1 inch in 60 minutes. In New York State the rate may be about 20% and in Pennsylvania about 30% higher. In Baltimore and Washington we may expect a 5-inch rate for 5 minutes, a 4-inch rate for 10 minutes, a 2.7-inch rate for 20 minutes, a 2-inch rate for 30 minutes, and 1.4 inches in 60 minutes. In New Orleans a 5.5-inch rate for 5 minutes, a 4.5-inch rate for 10 minutes, a 3-inch rate for 20 minutes, a 2.5-inch rate for 30 minutes, a 2-inch rate for 60 minutes or even more may be expected. In the central States a rate of 3.7 inches for 10 minutes, 2.8 inches for 20 minutes, 2.3 inches for 30 minutes, and 1.7 inches in 60 minutes may be expected. Further data, however, may require a change in any of these values.

The last rates in the table are given to show what downpours sometimes occur in certain sections. In 2 hours there fell at St. Kitts more than 12 inches; how much more could not be ascertained. The Palmetto gauge was swept away after registering 12 inches in 2 hours.

The records seem to show, where any information on the subject is given, that the maximum intensity usually lasts but a few minutes, seldom more than ten; that it sometimes occurs at the beginning of a storm, but in a great majority of instances occurs at the middle or end of it, quite a number stopping 10 to 20 minutes after the maximum rate is attained. As to the area simultaneously covered by the maximum rates of fall, almost no data are available.

Table No. 9.

MAXIMUM AMOUNTS OF RAIN FALLING DURING DIFFERENT PERIODS OF TIME.

\multicolumn{9}{c}{Length of Period in Minutes.}									
5	10	15	20	25	30	45	60	Over 60 Amt.	Duration	Place and Date.
....	....	....	....	....	....	....	0.50	5.20	24 h.	Boston, Oct. 12, 1895
....	1.40	....	1.50	....	....	....	.....	....	....	" { 1879–1891
....	0.80	....	1.30	....	....	....	.....	....	....	" July 18, 1884
....	0.70	....	....	....	....	....	.....	...	....	Providence, R. I., May 18, 1877
....	....	....	....	....	0.50	....	.....	....	....	" " Aug. 29, 1877
....	....	....	....	....	1.50	....	.....	....	....	" " " 6, 1878
....	....	0.75	....	....	....	....	2.56	....	....	" " " 28, 1882
....	0.90	....	....	....	....	....	1.74	....	....	Ithaca, N. Y., Aug. 4, 1892. Preceded by 5 hours of light rain
....	....	....	....	....	....	....	.....	5.03	1¼ h.	Mt. Carmel, N. Y., July 2, 1897
....	....	....	....	....	....	....	.....	1.73	1¼ h.	Morrisania, Oct. 30, 1860
....	....	....	0.50	....	....	....	.....	1.18	2 h.	New York City, Sept. 19, 1894
0.35	....	....	....	....	....	....	1.40	....	....	" " " Aug. 19, 1893
....	1.20	....	....	....	....	....	.....	....	....	" " " 7 times during 1869–1891
..	0.80	....	1.30	....	1.50	....	.....	....	....	" " " 3 " " " "
....	0.60	....	....	....	....	....	.....	....	....	" " " 5 " " " "
....	....	....	....	....	....	....	.....	3.60	24 h.	" " " May 4, 1893
....	....	....	....	....	....	....	.....	6.17	24 h.	" " " 1882 [sewer]
....	....	....	0.80	....	....	....	1.00	....	....	" " " July 6, 1896. (Gorged a
....	....	....	....	2.00	....	....	.....	....	....	Brooklyn, N. Y.
....	1.00	....	....	....	....	....	.....	....	....	Spring Mount, Pa., June 6, 1893
0.25	1.00	....	1.50	....	0.45	....	.....	....	....	" " during 1893 (4 storms)
....	1.00	1.28	1.60	....	....	....	.....	....	....	Philadelphia, Pa., 2 times during 1884–1891
....	0.80	....	1.30	....	....	....	.....	....	....	" " 3 " " " "
....	0.60	....	....	....	1.00	....	.....	....	....	" " 5 " " " "
....	0.40	....	....	....	0.06	....	.....	....	....	" " March, 1890
0.21	0.35	0.46	0.58	0.62	0.64	0.83	0.95	....	....	Baltimore, 1896
0.25	0.60	0.85	0.92	0.95	....	....	.....	....	....	
0.80	1.60	....	....	....	2.00	....	.....	....	....	Washington, D. C., 2 times during 1871–1891
....	0.80	....	1.30	....	1.50	....	.....	....	....	" " 5 " " " "
....	0.00	....	....	....	....	....	.....	....	....	" " 10 " " " "
0.52	....	0.75	....	1.78	....	2.00	.....	....	....	" " mean of many rains
0.30	0.85	1.20	1.48	....	2.05	2.70	3.40	6.00	2 h.	New Orleans, June 17, 1895
0.40	0.70	1.00	1.32	1.61	1.87	2.62	3.19	3.30	2 h.	" " Aug. 13, 1894
0.40	0.75	1.20	1.82	1.95	2.15	2.45	2.70	....	....	" " July 4, "
0.60	1.20	1.50	1.75	2.00	2.10	2.20	2.32	2.35	1½ h.	" " " 14, "
0.25	0.60	1.00	1.17	1.25	....	....	.....	....	....	" " Sept., 1889 (2 storms)
..	....	....	....	....	....	4.12	3.60	....	....	" " April 24, 1894. Preceded by 6 hours of light rain
0.35	0.65	0.90	1.15	1.40	1.55	1.78	1.88	....	....	
0.14	0.20	0.47	0.85	0.90	1.01	1.71	1.99	....	....	Jacksonville, Fla., U. S. Weather Bureau, 1896
0.30	0.61	0.70	0.92	1.30	1.57	1.15	1.80	....	....	
0.07	0.26	0.44	0.77	0.82	0.91	1.50	.....	....	....	
0.10	0.30	0.37	0.47	0.62	0.75	0.88	1.00	....	....	Galveston, Texas, 1896. U. S. Weather Bureau
0.30	0.45	0.57	0.65	0.77	0.90	1.10	1.23	....	....	
....	1.00	....	....	....	....	....	.....	....	....	Chicago, once during 1889–1891
....	0.80	....	....	....	....	....	.....	....	....	" 2 times " " "
....	0.60	....	....	....	....	....	.....	....	....	" 2 " " " "
....	....	1.10	....	....	....	....	.....	1.75	2 h.	Ohio Valley, July 16, 1896
....	....	....	....	....	....	....	.....	3.30	2 h.	St. Louis, May 14, 1891
0.28	0.58	0.88	1.13	1.23	1.38	....	.....	....	....	Cleveland, Ohio, 1896
0.04	0.19	0.49	0.79	1.04	1.31	1.73	.....	....	....	
0.15	0.45	0.85	1.00	1.07	1.14	....	.....	....	....	Detroit, Mich., " U. S. Weather Bureau
0.38	0.57	0.63	0.91	1.13	1.30	1.70	1.78	....	....	
0.25	0.50	0.75	0.95	0.99	1.02	....	.....	....	....	Little Rock, Ark., "
0.06	0.33	0.46	0.60	0.66	0.70	0.79	0.91	....	....	
0.35	....	....	....	....	....	....	0.82	....	....	San Diego, Cal., December, 1896
....	....	....	....	....	....	....	11.5+	....	....	Campo, Cal., August, 1891
....	....	....	....	....	....	....	8.8	....	....	Palmetto, Nev., " 1890
....	....	....	....	....	....	....	.....	36+	24 h.	Island of St. Kitts
....	....	....	....	....	....	....	.....	12+	2 h.	

ART. 18. RUN-OFF DATA.

The data concerning rates of run-off as compared with rainfall during the same time are very meagre. The total annual or monthly proportion of run-off has been ascertained in many different localities, but even this is for natural wooded surfaces or fields only. The number of careful gaugings in this country of rainfall within city limits and of contemporaneous sewer discharge from a known area probably does not exceed a half dozen.

One of the most recent, extensive, and scientific of such gaugings was that made at New Orleans in 1894-5 under the direction of the Engineering Committee on Drainage. Unfortunately for its general usefulness in the study of run-off problems the run-off measured probably included seepage from a soil lower than, and consequently more or less saturated by, the Mississippi River. The rainfall was recorded continuously at several points throughout the city, and several of the maximum rates are given in Table No. 9. A continuous record was also kept of the amount of water reaching the drainage-ditches from above and beneath the surface of the soil. From data thus and otherwise obtained the committee prepared curve-diagrams (Plate No. II) for the calculation of run-off from areas of different extent, character, and grade of surface in New Orleans. "The set marked A represents the run-off from densely built-up parts; the set marked B applies to the areas having small yards, or a medium density of population; the set marked C applies to the sparsely built-up parts, or those having large yards; and the set marked D applies to the rural areas. These curves, therefore, indicate the maximum rate of rainfall which it is proposed to provide for, and which is assumed to reach the drains and canals from the respective areas.

"They do not warrant the assumption, however, that the

discharge will never exceed the quantities given for it; in fact it is certain that they will be exceeded, but at such rare and indefinite intervals that their consideration is not justified. It should also be remarked that the curves are based upon the assumption that . . . the water enters the drains promptly, as is the case in most other cities." (Report of the Engineering Committee—B. M. Harrod, Henry B. Richardson, and Rudolph Hering—on the Drainage of the City of New Orleans; 1895.)

A number of gaugings have been made in Washington, D. C., in districts whose streets are almost entirely paved with asphalt. In one case "the flow in the sewer rose almost immediately after the rain began and fell to its normal level within a few minutes after the rain ceased." During another storm "at its maximum period the rain fell for 37 minutes at the rate of 0.9 of an inch per hour. The sewer-gauge rose to a height of 3.7 feet, giving about 0.47 of the capacity of the sewer and indicating no loss whatever by absorption or evaporation during the time of maximum flow." (Hoxie on "Excessive Rainfalls," Transactions Am. Soc. C. E., vol. xxv.) This sewer received the drainage of 200 acres.

Gaugings made in Rochester by Emil Kuichling are too extensive to be quoted here, but the tables may be found in the Transactions Am. Soc. C. E., vol. xx, pages 1–60, accompanied by an excellent discussion on the subject of run-off. The conclusions drawn from these by the author of that paper are quoted, as stating clearly the general principles on which are founded the rational methods of calculating run-off. The gaugings, he says, "point unmistakably to the following general conclusions:

"1. The percentage of the rainfall discharged from any given drainage-area is nearly constant for rains of all considerable intensity and lasting equal periods of time. This cir-

AMOUNT OF SEWAGE. 49

Plate II.
MAXIMUM FLOW FROM DRAINAGE AREA IN CUBIC FEET PER SECOND

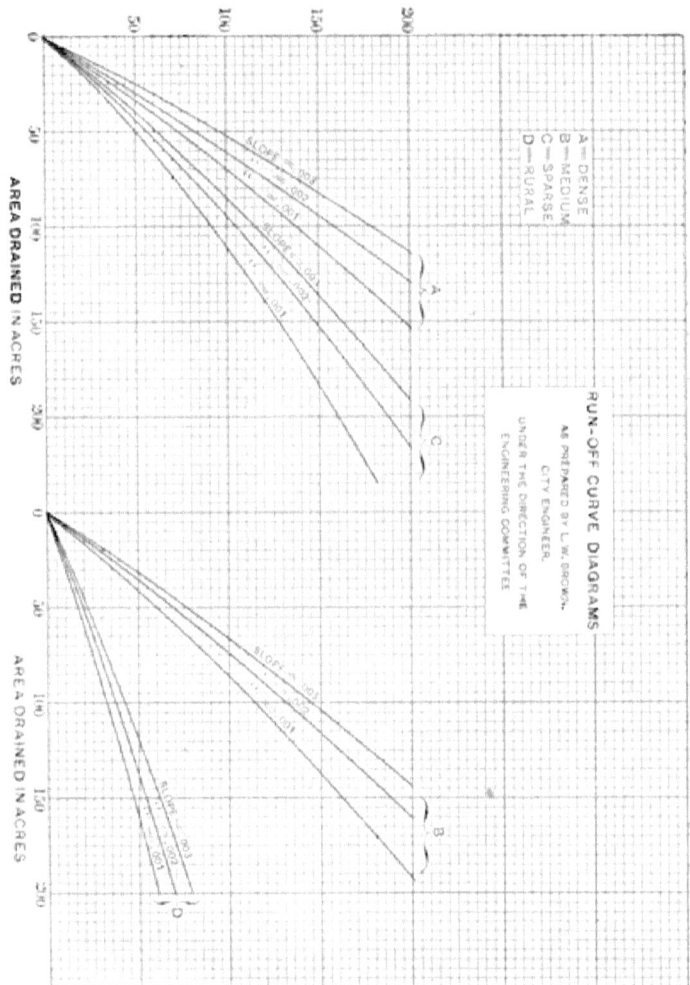

cumstance can be attributed only to the fact that the amount of impervious surface on a definite drainage-area is also practically constant during the time occupied by the experiments.

"2. The said percentage varies directly with the degree of urban development of the district, or, in other words, with the amount of impervious surface thereon. . . .

"3. The said percentage increases rapidly, and directly or uniformly with the duration of the maximum intensity of the rainfall, until a period is reached which is equal to the time required for the concentration of the drainage-waters from the entire tributary area at the point of observation; but if the rainfall continues at the same intensity for a long period, the said percentage will continue to increase for the additional interval of time at a much smaller rate than previously. This circumstance is manifestly attributable to the fact that the permeable surface is gradually becoming saturated and is beginning to shed some of the water falling upon it; or, in other words, the proportion of impervious surface slowly increases with the duration of the rainfall.

"4. The said percentage becomes larger when a moderate rain has immediately preceded a heavy shower, thereby partially saturating the permeable territory and correspondingly increasing the extent of impervious surface.

"5. The sewer discharge varies promptly with all appreciable fluctuations in the intensity of the rainfall and thus constitutes an exceedingly sensitive index of the rain and its variations of intensity.

"6. The diagrams also show that the time when the rate of increase in the said percentage of discharge changes abruptly from a high to a low figure agrees closely with the computed lengths of time required for the concentration of the storm-waters from the whole tributary area, and hence the said percentages at such times may be taken as the proportion of impervious surface upon the respective areas." (Transactions Am. Soc. C. E., vol. XX, page 37.)

"The Nagpoor (India) storage reservoir receives the flow from a watershed of 6.6 square miles. With a very absorbent natural surface that watershed has nevertheless delivered to the reservoir in 170 minutes 98% of a downpour upon its entire area of 2.2 inches in 80 minutes, when the power of absorption of the soil had been satisfied." (Hoxie).

Many instances could be named where storm-sewers which were designed to carry a run-off of one cubic foot per second have caused serious damage by their too small capacity; several where even a capacity of two cubic feet per second was insufficient. (One inch of rainfall per hour equals one cubic foot per second per acre almost exactly.) Not many years ago a sewer was considered by most engineers to be of ample size if it was designed for a rainfall of one inch per hour, one half running off; but the insufficiency of this rule has been learned by costly experience.

Accounts of accidents through insufficient sewer dimensions are unfortunately more numerous than data giving exact figures of unusual volumes of rainfall reaching sewers.

An analysis of the available data seems to point to the following conclusions:

That the total run-off from any area is directly proportional to the imperviousness of the surface, and that this imperviousness increases with the length of the storm, unless it is already 100%.

That very nearly 100% of the water falling upon an impervious surface flows immediately to the sewer unless held back by obstructions in the street, roof-gutters, or sewer-inlets—the last including insufficiency of size of the inlet. A small percentage, however, is usually evaporated at once.

That the proportion of the rainfall on any given impervious area which reaches any particular point in the sewer system increases with the length of the storm up to the time when the run-off from the most distant part of said area reaches the point of observation; after which the run-off very nearly equals the rainfall upon said area while the rate of fall remains constant.

That the percentage of imperviousness of the surface may vary from 0% to 100%, being the first in the case of very porous soil under natural conditions at the beginning of a rain, and the last in an urban district where streets, sidewalks, and yards are all paved, or occasionally where a dense clay soil is saturated by previous rainfall.

ART. 19. FORMULAS FOR STORM-WATER RUN-OFF.

Many attempts have been made to construct a simple general formula for obtaining the run-off from any area. The best known of these are as follows:

Craig: $D = 440BN \text{ hyp. log } \frac{8L^2}{B}$.

$D =$ discharge in cubic feet per second;
$L =$ extreme length of drainage-area;
$B =$ mean breadth of drainage-area;
$N =$ constant varying from 0.37 to 1.95.

Dredge: $Q = 1300 \frac{M}{L^{\frac{1}{3}}}$.

$L =$ length of watershed;
$M =$ area in square miles.

Dickens: $D = 825 M^{\frac{3}{4}}$.

$D =$ discharge in cubic feet per second;
$M =$ drainage-area in square miles.

Fanning: $Q = 200 M^{\frac{5}{6}}$.

$Q =$ discharge in cubic feet per second;
$M =$ drainage-area in square miles.

Bürkli-Ziegler: $Q = Rc\sqrt[4]{\dfrac{S}{A}}$.

c = constant—0.75 for paved streets, 0.31 for macadamized streets;

R = average rate during heaviest fall in cubic feet per second per acre;

S = general fall of area per 1000;

Q = cubic feet per second per acre reaching sewers;

A = drainage-area in acres.

Kirkwood: $D = \left(\dfrac{N^2}{5804 S}\right)^{\!6}$.

D = diameter of sewer in feet;

S = sine of inclination;

N = number of acres in area.

Maximum rainfall of one inch, one half running off.

Hawksley: $\log D = \dfrac{3 \log A + \log N + 6.8}{10}$.

D = diameter of sewer in inches;

A = number of acres to be drained;

N = length in feet in which sewer falls one foot.

City or suburban surfaces.

Adams: $\log D = \dfrac{2 \log A + \log N - 3.79}{6}$.

A = area in acres;

N = as in Hawksley;

D = diameter in feet.

For one inch rainfall, one half running off.

McMath: $Q = Rc\sqrt[6]{\dfrac{S}{A}}$.

Terms as in Bürkli-Ziegler; R taken at St. Louis as 2.75 inches.

Kuichling: $Q = Aat(b - ct)$.

Q = discharge in cubic feet per second;
A = drainage-area in acres;
t = duration in minutes of the intensity $(b - ct)$;
$b = 2.1$ } for Rochester, N. Y. (Kuich-
$c = 0.0205$ } ling recently gives as a formula representing storms of the second class at Rochester:

$$r = \frac{12}{t^{.06}},$$ in which r = rate in inches per hour.)

$$a = \frac{\text{proportion of impervious surface}}{t}.$$

The following, known as Roe's tables, gives the number of acres of urban surfaces which can be drained by sewers of different diameters and at different grades. It is no longer in general use.

Inclination, Fall, or Slope of Sewer.	Inner Diameter or Bore of Sewer in Feet.								
	2	2½	3	4	5	6	7	8	9	10
Level.	39	67	120	277	570	1020	1725	2850	4125	5825
¼ in. in 10 ft., or 1:480....	43	75	135	308	630	1117	1925	3025	4425	6250
½ " " " " " 1:240....	50	87	155	355	735	1318	2225	3500	5100	7175
¾ " " " " " 1:160....	63	113	203	460	950	1692	2875	4500	6575	9250
1 " " " " " 1:120....	78	143	257	590	1200	2180	3700	5825	7850	11,050
1½" " " " " 1:80.....	90	165	295	670	1385	2486	4225	6625	....	....
2 " " " " " 1:60.....	115	182	318	730	1500	2675	4550	7125	....	....

The formulas of Craig, Dredge, Dickens, and Fanning apply to natural surfaces and have the shape and extent of the drainage-area as the only variables. The Bürkli-Ziegler, Kirkwood, and McMath formulas take into account also the slope of the surface. The Bürkli-Ziegler and Kuichling allow for varying conditions of surface, and, together with the McMath, for varying rates of rainfall. All these formulas except the three last mentioned are based on an assumed maximum rate of rainfall.

Roe's tables give the diameter of sewer necessary to meet various conditions of area and sewer grade. As in a level sewer the surface of the water must have some fall if there is to be any flow, the quantities given for a level grade can apply only to a limited length of sewer. None of these formulas is satisfactory for all cases, because none takes into account all the variable conditions. Those which are probably the most frequently used are the Bürkli-Ziegler, McMath, and Kuichling, and these are seen to be the ones containing the most variables. The proper test of any formula is to calculate by it from known data quantities which are also known. Many such tests of all these formulas have been made, and it has been found that there are few, if any, cases in which all will give results practically identical or equal to the actual quantities as measured. Such a comparison is given of Roe's tables, Hawksley's, Kirkwood's, and Bürkli-Ziegler's formulas, and the actual gaugings of a sewer in Washington made in 1884 (from paper by Capt. R. L. Hoxie read before the Am. Soc. C. E. July 2, 1886):

Rainfall.	Roe's Tables.	Hawksley.	Kirkwood.	Bürkli-Ziegler.	Actual Maximum Flow.
0.5" in 15 min....	36.3	43.2	51.7	137.6	300
0.55" in 37 min..	36.3	43.2	51.7	61.9	180

The discrepancies are largely due to the causes already referred to—that factors are taken as constants which are really variables, and hence each formula can give correct results for certain cases only. In most the constant is supposed to be derived from maximum rates of rainfall, but such data were until recently incomplete and inaccurate. Also, since the authors of the older formulas were Europeans or derived their data from European sources, the maximums were those for Europe and are not applicable to this country. Also the character of the majority of city-street surfaces has changed since that time. The Kuichling, Bürkli-Ziegler, and McMath formulas recognize the variableness in drainage-surfaces.

It is possible that a formula can be devised which shall represent by variable factors all the conditions which have been shown to affect the run-off. But it can hardly be expected that such a formula can be other than cumbersome; and it is probable that the shortest method which is at all rational and accurate in all cases is that of subdividing the calculation, and adapting a general method rather than a general formula to the peculiar conditions of each case. Such a method is recommended and will be outlined further on.

Many engineers, however, use some one of the formulas given, and a large majority of the storm-sewers built in this country are probably so designed—in spite of the fact that McMath considers his formula (which is probably the most popular) as adapted to large areas only, and that it is derived in an entirely empirical manner from St. Louis data only; and that Kuichling has "finally abandoned the attempt to establish a general formula for run-off," although the one bearing his name is largely general in application and rational in origin and construction.

ART. 20. EXPEDIENCY OF PROVIDING FOR EXCESSIVE STORMS.

An examination of rainfall records shows no apparent law of frequency of excessive storms. It can be said as a general statement, however, that a rate of fall within certain limits may be expected almost any month; one within higher limits five or more times in ten years (these are the storms referred to in Art. 16); and a phenomenal downpour at most irregular intervals, usually many years apart. Should the sewer be designed to carry the run-off from storms of the first class only, of the second class, or be of the greatest size demanded by the third class?

That the last is desirable will not be disputed, but both practical and financial difficulties frequently oppose this course. The practical ones, however, in most, if not all, cases resolve themselves into financial ones; and the question becomes one of dollars and cents, and to a certain extent also of public convenience, which cannot be assigned a money value. To accommodate the second class of storms may require a sewer of three or four times the capacity of one which would suffice for the run-off from the first class, and the third class a capacity two or three times that of the second. The result of providing capacity for the first class only would probably be flooding of streets and cellars one or more times almost every year; for the second class, flooding at intervals of several years; for the third class, perfect immunity from floods.

The loss resulting from a flooding of streets and cellars by water more or less foul may be very considerable; goods may be damaged, business suspended, foundations weakened, health threatened by the dampness lingering after the floods have withdrawn; also real-estate values in a district liable to floods will depreciate and the city as a whole will be a loser by increased tax rates to meet the decreased valuation. On the other hand as the capacity of a sewer increases so does the cost, and this fact may place an urgent, even an imperative, limit to either the capacity or the extent of the sewers to be built. The relative cost of sewers of different capacities (other things supposedly equal) in Washington, D. C., for about 10 years is given in the following table (adapted from one prepared by Capt. Hoxie). The unit of capacity is that of a 12-inch pipe.

The exact proportion between the capacity and the cost, and the rule of their relative increase, will vary with different

Table No. 10.

Number of Units of Capacity	Relative Cost per Foot.	Number of Units of Capacity.	Relative Cost per Foot.	Number of Units of Capacity.	Relative Cost per Foot.
1	1.000	6	1.920	20	3.170
2	1.174	7	2.090	30	3.480
3	1.388	8	2.250	40	4.030
4	1.567	9	2.410	50	4.170
5	1.743	10	2.570		

methods of construction, depth of sewer, etc.; and in many localities the cost will be found to increase more rapidly relative to the capacity than is indicated by Table No. 10. But in any case it will be found that the increase of cost is much less rapid than that of capacity. Using the table of cost of Washington sewers, a sewer of three times the capacity of a 12-inch pipe would cost 1.388 units of value. If this would just suffice for the heaviest storms of the first class on a given area one of a capacity ample for the maximum of the second class, or four times as great, would cost 2.8 units of value, or about twice as much; and one capable of removing the run-off from the greatest downpours, or twelve times capacity of the first, would cost 3.75 units of value, or only 2.7 times as much as the first. Moreover, as we shall see later, the larger mains need to increase less rapidly in capacity, and hence in cost, to accommodate the heavier storms than do the smaller laterals used in this illustration.

The decision as to which class of storms the size shall be adapted to must be made for each case by the engineer or the city authorities as their judgment dictates. But probably in nine cases out of ten the truest economy will be observed by constructing sewers sufficient for the second, or even in some cases the third, class of storms. The damage likely to result from the use of sewers of too small capacity, which damage is to be balanced against the extra cost of larger ones, will depend upon circumstances. "If the surface flow upon the streets passes off to a proper outlet without causing damage or inconvenience the flood is well disposed of. If not, there is danger in permitting stormwater to accumulate upon streets with steep grades. It becomes a torrent flowing with great velocity, and cannot then be captured by inlets designed to arrest, each, its share of shallow gutter flow with small velocity. It moves rapidly down to valleys or basins without surface outlet; here it floods the surface, because inlets to receive it as fast as it comes can rarely be constructed—even should the drains here be of sufficient capacity. Inlets for large volumes of water in city streets are apt to be pitfalls for pedestrians and traps for cart-wheels and horses' feet. If the drains of the inundated district are of insufficient capacity the consquences are, of course, disastrous." (Hoxie, " Excessive Rainfalls.")

As suggested in this quotation, it is possible in many localities to lead the surplus water of severe storms over the surface to the nearest natural watercourse, and this without any damage resulting, although it may be temporarily inconvenient. But in a city where all small streams have been walled in or diverted to sewers this is possible only along a water front.

CHAPTER IV.

FLOW IN SEWERS.

Art. 21. Fundamental Theories.

The flow in an ordinary sewer must be due to one cause only—the attraction of gravitation. The velocity of this flow is retarded by friction and other obstacles affecting it along the line of the sewer.

The general formula for the velocity due to gravity of a freely falling body is $V = \sqrt{2gh}$, where V is velocity in feet per second, h is head in feet, and g is acceleration due to gravity, being about 32.16 feet per second. In the case of running water h is the fall of the surface of the water from the point of no motion to the point in question. Therefore if there were no opposing forces a stream would flow more and more rapidly along its course as the total head became greater; and its velocity would become constant only when the surface was level, and therefore h constant. There is, however, friction between the moving water and the sides of a sewer, and this must be overcome by some force. Since the only force available is that due to gravity, called into play by the creation of a head h, a part of this force must be used in overcoming friction. If it is not all so used the remainder goes to create additional velocity. Friction, it is found, increases with the velocity of the moving body, so that, as additional increments of speed are created by h, a larger proportion of the head is consumed in overcoming friction, until at last all of h is so consumed and none goes to increasing the velocity—that is, the velocity remains constant. Friction also varies with the roughness of the surface. The total amount of energy lost in friction also increases with the duration of its action, which is proportional to the distance travelled l. It is found that at least one other condition affects the amount of friction in sewers, viz., the proportion between the cross-sectional area of the stream and the length in this cross-section of the line of contact between the water and the bed of the stream; the greater the first is in proportion to the second the less the effect of friction upon the mass as a whole. This proportion, or $\dfrac{\text{area of section}}{\text{wetted perimeter}}$, is customarily represented by R and is called "hydraulic radius" or "mean depth."

From these considerations it follows that V varies as $\sqrt{2gh}$ and as $f(R)$, and inversely as $f(l)$. The effect of roughness may be represented by a factor a. A formula for velocity would therefore be in the form $V = a\dfrac{\sqrt{2gh}f(R)}{f(l)}$. In 1753 Brahms proposed as a formula representing the resultant effect of these accelerating and retarding influences $V = c\sqrt{RS}$, in which $V =$ mean velocity of current, c is an empirical constant which includes $\sqrt{2g}$ and a, R is the hydraulic radius, and S is the sine of the surface slope, or $\dfrac{h}{l}$. This formula, now generally called Chézy's formula, has been made the basis of others, most of which differ among themselves only in the values given to c; but it is now recognized that V does not vary exactly as the square root of R and of S; that is, that $f(R)$ and $f(l)$ in the formula $V = a\dfrac{\sqrt{2gh}f(R)}{f(l)}$ are not exactly $\sqrt{R}$ and $\sqrt{l}$; but they approximate it, and this formula may therefore be written

$$V = a\sqrt{2g}\frac{f'(R)}{f'(l)}\sqrt{\frac{h}{l}R} = b\frac{f'(R)}{f'(l)}\sqrt{RS},$$

$b\dfrac{f'(R)}{f'(l)}$ being equal to the c of Chézy's formula. From this it follows that c is not a constant for any particular sewer or stream, but varies with both R and S. The principal cause affecting the value of c, however, is the condition as to roughness of the wetted perimeter.

If we wish to obtain the velocity of flow in any sewer by this formula it is necessary to select proper values for c, R, and S. S can be readily obtained by dividing h by l. The value of c and R and their relation to V will now be discussed.

For c most of the older formulas give constant values; but since V varies with different materials of channel-walls, whose character does not affect the values of R and S, this variation must be recognized in a variable c by means of a new factor or by a new equation. Most of the efforts looking to greater accuracy have been directed toward determining values for c and thousands of experiments have been made for this purpose. D'Arcy's value, somewhat simplified and for feet measure, is

$$c = \left(\dfrac{155256}{12D + 1}\right)^{\frac{1}{2}}.$$

Bazin's value for cut stone and brick-work is

$$c = \left(\dfrac{1}{.0000133\left(4.354 + \dfrac{1}{R}\right)}\right)^{\frac{1}{2}}.$$

Eytelwein's value is $c = 93.4$.

The formula evolved from the records of a large number of experiments by Messrs. Ganguillet and Kutter, usually called "Kutter's formula," is now generally held to give results more nearly approximating the actual velocities than any other. This formula is, for English measure,

$$c = \dfrac{41.6 + \dfrac{.00281}{S} + \dfrac{1.811}{n}}{1 + \left(41.6 + \dfrac{.00281}{S}\right)\dfrac{n}{\sqrt{R}}},$$

in which n is a "coefficient of roughness" of the sides of the channel, such coefficient having been obtained by averaging many experiments. In the selection of value for n great care and judgment must be exercised, particularly for small sewers, in the calculation for which n has a greater effect than in that for large channels.

The values of n are approximately:

Sides and bottom of channel lined with well-planed timber. .009
With neat cement, clean glazed sewer-pipe, and very
    smooth iron pipe............................... .010
With 1:3 cement mortar or smooth iron pipe........... .011
With unplaned timber and ordinary iron pipe........... .012
With smooth brick-work or ordinary pipe sewers........ .013
With ordinary brick-work .... .015
With rubble or granite-block paving................... .017

Kutter's formula is seen to provide for variations in c due not only to the character of the channel but also to changes in R and S.

This formula has been used to calculate the tables Nos. 11 and 12, n being taken as .013 in the former and .015 in the latter. If it is desired to use another value of n the corresponding values of velocity and discharge can be obtained very approximately by multiplying the quantities given in each table by the factors given below it for that purpose. For ordinary pipe or good brick sewers n may be taken as .013, for ordinary brick or smooth stone as .015. For extra smooth work n may be taken as .011.

The uncertainties necessarily existing in the estimates of the amount of sewage to be provided for and the difficulty of selecting just the proper value for n, owing to the non-uniform character of the interior surface of the sewer, make a refinement of calculations out of keeping with the data used. Moreover, in the case of vitrified clay or concrete pipe the

Table No. 11.

VELOCITY AND DISCHARGE IN SEWERS 4 TO 36 INCHES DIAMETER.

Velocity in Feet per Second; Discharge in Cubic Feet per Minute; Sewers Flowing Full.

(Formula $V = c \sqrt{RS}$; c calculated by Kutter's formula, with $n = .013$. $Q = 60aV$.)

Grade of Sewer.	4-inch		6-inch		8-inch		10-inch		12-inch		15-inch		18-inch	
	V	Q	V	Q	V	Q	V	Q	V	Q	V	Q	V	Q
.1	5.75	30.13	7.99	94.10	10.04	210.3	11.94	390.8	13.73	647.0	16.24	1196.0	18.59	1971.5
.05	4.06	21.28	5.64	66.48	7.09	148.6	8.43	276.1	9.70	457.1	11.48	845.0	13.13	1393.0
.04	3.63	19.03	5.05	59.45	6.34	132.9	7.54	246.9	8.65	407.8	10.26	755.6	11.74	1244.0
.03	3.15	16.47	4.25	50.01	5.49	115.0	6.53	213.7	7.51	353.9	8.89	654.4	10.17	1078.0
.02	2.57	13.44	3.56	42.00	4.48	93.90	5.33	174.5	6.13	289.0	7.25	534.2	8.30	880.4
.01	1.82	9.50	2.52	29.70	3.17	66.38	3.77	123.4	4.33	204.3	5.13	377.8	5.87	622.6
.008	1.61	8.37	2.25	26.53	2.83	59.35	3.37	110.3	3.87	182.6	4.59	337.7	5.25	556.8
.006	1.38	7.18	1.95	23.00	2.45	51.39	2.92	95.55	3.35	158.2	3.97	292.5	4.55	482.3
.004			1.59	18.71	2.00	41.85	2.38	77.98	2.74	129.1	3.24	238.3	3.70	392.9
.002					1.40	29.38	1.67	54.61	1.91	90.40	2.27	167.3	2.60	275.9
.001							1.17	38.41	1.35	63.58	1.60	117.7	1.83	194.3
.0009											1.51	111.4	1.73	183.9
.0008													1.63	173.1
.0007													1.52	161.2

For $n =$	.011	.012	.013	.015	.017
Multiply V or Q by	1.20	1.09	1.00	0.84	0.73

Table No. 11.—Continued.

VELOCITY AND DISCHARGE IN SEWERS 4 TO 36 INCHES DIAMETER.

Velocity in Feet per Second; Discharge in Cubic Feet per Minute, Sewers Flowing Full.

(Formula $V = c \sqrt{RS}$; c calculated by Kutter's formula, with $n = .013$. $Q = 60aV$.)

Grade of Sewer.	20-inch		22-inch		24-inch		30-inch		33-inch		36-inch	
	V	Q	V	Q	V	Q	V	Q	V	Q	V	Q
.1	20.08	2628	21.51	3407	22.91	4319	26.84	7905	28.69	10220	30.46	12920
.05	14.18	1857	15.20	2407	16.19	3052	18.97	5586	20.27	7225	21.54	9136
.04	12.69	1661	13.59	2153	14.47	2729	16.96	4995	18.13	6461	19.26	8171
.03	10.98	1438	11.77	1864	12.53	2363	14.69	4325	15.70	5595	16.68	7075
.02	8.97	1174	9.61	1522	10.23	1930	11.99	3532	12.82	4568	13.62	5777
.01	6.34	830	6.79	1076	7.24	1396	8.48	2497	9.06	3230	9.63	4085
.008	5.67	742	6.07	962	6.47	1220	7.58	2233	8.11	2889	8.61	3653
.006	4.91	643	5.26	833	5.60	1057	6.57	1934	7.02	2502	7.46	3164
.004	4.00	524	4.29	679	4.56	860	5.35	1576	5.72	2040	6.08	2580
.002	2.81	368	3.01	477	3.21	605	3.76	1109	4.02	1434	4.28	1814
.001	1.98	259	2.12	336	2.26	427	2.66	782	2.84	1012	3.02	1281
.0009	1.87	245	2.01	318	2.14	404	2.51	741	2.69	959	2.86	1213
.0008	1.76	231	1.89	299	2.02	380	2.37	697	2.53	902	2.69	1141
.0007	1.64	215	1.76	279	1.88	354	2.20	650	2.36	841	2.51	1065
.0006	1.51	198	1.63	258	1.73	327	2.04	600	2.18	777	2.32	984
.0005			1.48	234	1.58	298	1.86	546	1.99	708	2.11	896
.0004			1.32	208	1.40	265	1.65	486	1.77	630	1.88	798
.0003					1.20	227	1.40	413	1.52	541	1.62	686
.0002					0.96	186	1.13	335	1.22	435	1.30	552

Table No. 12.

VELOCITY AND DISCHARGE IN SEWERS 33 INCHES TO 10 FEET DIAMETER.

Velocity in Feet per Second; Discharge in Cubic Feet per Minute; Sewers Flowing Full.

(Formula $V = c\sqrt{RS}$; c calculated by Kutter's formula, with $n = .015$. $Q = 60aV$.)

Grade of Sewer.	33-inch		36-inch		42-inch		4-foot	
	V	Q	V	Q	V	Q	V	Q
.05	17.17	6120	18.27	7750	20.37	11765	22.36	16865
.04	15.36	5473	16.34	6930	18.21	10517	20.00	15080
.03	13.30	4738	14.15	6000	15.77	9108	13.31	13057
.02	10.85	3568	11.55	4900	12.88	7437	14.13	10658
.01	7.68	2735	8.16	3464	8.90	5258	9.99	7537
.008	6.86	2444	7.30	3096	8.14	4700	8.93	6738
.006	5.94	2115	6.32	2679	7.04	4067	7.73	5832
.004	4.84	1726	5.15	2186	5.75	3243	6.31	4759
.002	3.41	1216	3.63	1540	4.05	2339	4.45	3354
.001	2.40	856	2.52	1085	2.85	1648	3.13	2365
.0009	2.27	810	2.42	1027	2.70	1561	2.97	2240
.0008	2.14	763	2.28	967	2.55	1470	2.80	2110
.0007	2.00	713	2.13	903	2.38	1373	2.61	1972
.0006	1.85	658	1.97	834	2.20	1269	2.42	1822
.0005	1.68	598	1.79	759	1.95	1128	2.20	1658
.0004	1.49	532	1.59	675	1.78	1028	1.96	1477
.0003	1.28	457	1.37	580	1.53	883	1.68	1270
.0002					1.23	712	1.36	1026
.00015							1.16	878

For $n =$ .011 .012 .013 .015 .017
Multiply V or Q by 1.43 1.29 1.19 1.00 0.87

TABLE No 12.—Continued.
VELOCITY AND DISCHARGE IN SEWERS 33 INCHES TO 10 FEET DIAMETER.

Velocity in Feet per Second; Discharge in Cubic Feet per Minute; Sewers Flowing Full.

(Formula $V = c\sqrt{RS}$; c calculated by Kutter's formula, with $n = .015$. $Q = 60aV$.)

Grade of Sewer.	5-foot		6-foot		8-foot		10-foot	
	V	Q	V	Q	V	Q	V	Q
.05	26.05	30700						
.04	23.30	27450	26.34	44690				
.03	20.17	23765	22.81	38700				
.02	16.47	19405	18.62	31600	22.53	67965	26.03	122700
.01	11.64	13717	13.17	22345	15.93	48050	18.41	86755
.008	10.41	12267	11.78	19980	14.25	42970	16.46	77590
.006	9.01	10617	10.19	17295	12.33	37200	14.25	67175
.004	7.36	8665	8.32	14113	10.07	30370	11.63	54840
.002	5.19	6110	5.87	9956	7.10	21435	8.21	38720
.001	3.66	4311	4.14	7030	5.02	15150	5.81	27380
.0009	3.47	4083	3.92	6659	4.76	14353	5.51	25960
.0008	3.27	3849	3.70	6276	4.49	13533	5.19	24475
.0007	3.05	3597	3.46	5870	4.20	12660	4.86	22895
.0006	2.82	3326	3.20	5429	3.88	11710	4.50	21195
.0005	2.57	3028	2.92	4946	3.54	10675	4.10	19325
.0004	2.29	2700	2.60	4411	3.16	9532	3.66	17267
.0003	1.97	2324	2.24	3801	2.73	8228	3.17	14927
.0002	1.60	1882	1.82	3083	2.22	6694	2.58	12168
.00015	1.37	1615	1.56	2650	1.93	5820	2.23	10510
.00012			1.39	2353	1.70	5137	1.99	9375
.00010					1.55	4672	1.81	8542
.000095					1.25	3783	1.77	8320
.000090							1.72	8096

market sizes must in the end be those selected, and there is a considerable jump between the capacities of consecutive sizes. For instance, an 8-inch pipe on a 1% grade will discharge about 498 gallons per minute when running full; a 10-inch pipe running full with the same grade will discharge about 925 gallons per minute, and a 12-inch pipe about 1530 gallons per minute. For this reason it is sufficiently accurate and often more convenient to use curves plotted from the tables, having the grade and corresponding velocity or discharge as coördinates, from which the flow through any customary size of sewer at any practicable grade can be found at a glance and with as great accuracy as is required for ordinary use. Such a diagram can be readily prepared on a sheet of cross-section paper, a curve being drawn for the velocity and another for the discharge of each size of sewer.

It is now generally considered that Kutter's formula gives somewhat too small values for sewers under 15 or 18 inches diameter.

It must be remembered that the formulas and tables of velocity are supposed to apply only when the sewage has reached a constant velocity. Previous to this when the friction does not consume all of h the remainder is creating increments of velocity. Since the same amount of sewage must pass all sections of a sewer between two inlets, however, it follows that, previous to the flow obtaining its maximum and constant velocity, the depth of sewage must have been greater, increasing up stream to the point of entry. An initial velocity of entrance in the direction of the sewage flow will reduce the amount and extent of this non-uniform flow with larger cross-section, but will have little effect upon the ultimate constant velocity. If no such initial velocity exist the entering sewage must, if it be any large percentage of the capacity of the sewer, back up the feeding-pipe through which it entered in order to create additional head h.

V is the mean velocity. The effect of friction is exerted along the wetted perimeter and grows less toward the centre of the stream. The surface of flow is also retarded by friction with the air, and frequently in the case of house-sewage by a greasy scum which floats upon the surface. The velocity given is really the volume of flow divided by its area.

Since V varies as $f(R) = f\left(\dfrac{\text{area}}{\text{wetted perimeter}}\right)$, it follows that the size of the sewer and the shape of the cross-section have considerable effect upon the velocity of a stream. The maximum value of $\dfrac{\text{area}}{\text{perimeter}}$ for a sewer flowing full is obtained, we learn from geometry, by making the cross-section circular; that is, for pipes of the same area, but different shapes of cross-section, flowing full, the circular gives the largest R. But this is not generally true when the sewer is *not* flowing full.

If we examine the effect of depth of flow in a given circular sewer upon the value of R we find that if the depth $d = \dfrac{D}{2}$ (D equalling the diameter of the sewer) $a = .3927 D^2$, $p = 1.5708 D$, and $R = 0.25 D$. If the depth $= D$ we find $a = 0.7854 D^2$, $p = 3.1416 D$, and $R = 0.25 D$ as before.

TABLE No. 13.

d Depth.	p Wetted Perimeter.	a Area of Flow.	R Hydraulic Radius.	$\sqrt{R}$	By Kutter's Formula.	
					Corrected Proportional Velocities.	Corrected Proportional Discharge.
Full. 1.0	3.142	0.7854	0.250	1.00	1.00	1.000
0.95	2.691	0.7708	0.286	1.07	1.11	1.068
0.9	2.498	0.7445	0.298	1.09	1.15	1.073
0.8	2.214	0.6735	0.304	1.10	1.16	0.98
0.7	1.983	0.5874	0.296	1.08	1.14	0.84
0.6	1.772	0.4920	0.278	1.05	1.08	0.67
0.5	1.571	0.3927	0.250	1.00	1.00	0.50
0.4	1.369	0.2934	0.214	0.93	0.88	0.33
0.3	1.159	0.1981	0.171	0.83	0.72	0.19
0.25	1.047	0.1536	0.146	0.76	0.65	0.14
0.2	0.927	0.1118	0.121	0.69	0.56	0.09
0.1	0.643	0.0408	0.0635	0.50	0.36	0.03

As the depth of the sewage decreases from that of half the diameter the area decreases more rapidly than does the wetted perimeter, and consequently R decreases more and more rapidly as the depth diminishes. The above table shows this very plainly. The diameter is here taken as unity, the sewer circular.

The formula for R for circular sewers for any given depth of flow is

$$R = \frac{\text{area}}{\text{wetted perimeter}} = \frac{\frac{2a\pi r^2}{360} - r^2 \sin a \cos a}{\frac{2a}{360} \times 2\pi r},$$

$$= \frac{r}{2}\left(1 - \frac{180 \sin a \cos a}{a\pi}\right),$$

in which $r =$ the radius of the sewer perimeter;

$a =$ the number of degrees in the angle whose cosine is $\dfrac{r - \text{the depth of flow}}{r}$.

For the egg-shaped sewer (see Art. 24) somewhat different values are found.

TABLE No. 14.
EGG-SHAPED SEWER.

($D =$ horizontal diameter; $H =$ vertical diameter.)

d in parts of H.	d in parts of D.	p in parts of D.	a in parts of D^2.	R in parts of D.	$1.8587\sqrt{R}$	By Kutter's Formula.		d in Circular Sewer in parts of D.[*]
						Corrected Proportional Velocities.	Corrected Proportional Discharge.	
Full 1.000	1.50	3.965	1.1485	0.2897	1.000	1.00	1.00	1.209
0.667	1.00	2.394	0.7558	0.3157	1.045	1.06	0.69	0.750
0.333	0.50	1.374	0.2840	0.2066	0.846	0.77	0.18	0.354
0.267	0.40	1.159	0.20485	0.1768	0.781	0.70	0.12	0.284
0.220	0.33	1.012	0.15510	0.1532	0.727	0.63	0.081	0.228
0.200	0.30	0.937	0.13471	0.1437	0.704	0.60	0.064	0.214
0.133	0.20	0.706	0.07497	0.1062	0.606	0.49	0.030	0.141
0.067	0.10	0.463	0.0279	0.06026	0.455	0.33	0.008	0.075
0.033	0.05	0.321	0.0102	0.03177	0.331	0.23	0.002	0.039

[*] To give equal discharge in circular sewer of same capacity—i.e., one whose diameter = 1.209 D.

By Table No. 13 it is seen that when a circular sewer is half full the wetted perimeter and area of flow are each half of that for a full sewer. When the depth is but $\frac{1}{4}$ the diameter, however, the wetted perimeter is $\frac{1}{3}$, the area of flow less than $\frac{1}{5}$, and R about $\frac{4}{7}$ that of a full sewer; and when the depth is $\frac{1}{10}$ the wetted perimeter is about $\frac{1}{5}$, the area $\frac{1}{19}$, and R about $\frac{1}{4}$ that of a full sewer.

In the last two columns we have the proportional velocities and discharges for various depths of flow, with allowance made for variations in c, calculated by Kutter's formula, with sufficient accuracy for ordinary use. The fifth column shows proportional velocities if c is considered as not affected by changes in R. A comparison of the fifth and sixth columns shows the effect upon the coefficient c of variations in R, since if $c = x$ for a full sewer for one .2 full it equals $\frac{89}{83}x$ and for one .8 full $\frac{118}{103}x$.

Reference to Table No. 13 shows that if, in a circular sewer with a depth of flow of $\frac{1}{4}$ the diameter, the velocity is $1\frac{1}{2}$ feet per second (the minimum velocity of flow ordinarily permissible for house-sewers), in the same sewer flowing full the velocity will be 2.3 feet per second. It also appears from this table that the greatest velocity is attained, not when the sewer is flowing full, but when the depth is .81 of the diameter, and that the maximum discharge occurs when the depth is .9 of the diameter. From this it follows that a circular sewer can never flow full unless under a head.

The tables Nos. 11 and 12 for flow in sewers give the velocity and discharge for full sewers only, the velocity being the same for a sewer half full, while the discharge is one half as great. They do not give the maximum capacity of the sewer, which is theoretically 1.07 times that given; but the velocity and discharge for sewers flowing full are most convenient for use and are on the safe side of exact accuracy.

Where it is desired to obtain the velocity or discharge of

a sewer flowing partly full the tables can be entered with the quantities corresponding to the other conditions, the velocity or discharge of the sewer as if it were flowing full obtained, and such part of this taken as is indicated by the above table for the given depth. For instance, if it is desired to find the discharge of a 10-inch circular sewer, grade 1 : 200, when the depth of flow is 0.4 the diameter, we find from the table that the discharge if running full would be about 650 gallons per minute; we multiply this by 0.33 and obtain 214 gallons, the volume required. Or, given the volume, 215 gallons, and the grade, 1 : 200, to find the depth of flow: we find the flow of a full sewer, 650 gallons, divide 215 gallons by this, obtaining $\frac{1}{3}$, and find the depth corresponding to this proportion of the discharge, or 0.4.

The velocity obtained by the formula or from the table is that for a straight pipe of a uniform cross-section and condition of surface. In a system of sewers there are numerous curves, irregularities of surface, manholes, house-branches, etc., each of which may exert a retarding influence upon the sewage. It is thought that there is no appreciable diminution of velocities in a curve whose radius is at least 5 times the diameter of the sewer. Weisbach's formula for loss of head in curves is

$$h = \frac{caV^2}{11578},$$

in which $c = .131 + 1.847\left(\frac{r}{b}\right)^{\frac{1}{2}}$;

$r =$ radius of pipe;
$b =$ " " bend;
$a =$ angle in degrees;
$V =$ velocity in feet per second;
$h =$ head in feet necessary to overcome resistance of curve.

From the above formula we find that if

$\dfrac{r}{b} =$ .1   .2   .3   .4   .5   .6   .7   .8   .9   1.0

then

$c =$ .131 .138 .158 .206 .294 .440 .661 .977 1.408 1.978

As an example, assume a 10-inch pipe, or $r = 0.42$ feet, that $b = 2$ feet, that $a = 90°$, that $V = 3$ feet. Then $h = \dfrac{.138 \times 90 \times 9}{11578} = .0097$ feet, or less than $\frac{1}{8}$ inch. This result is not sufficiently large to materially affect the design. It represents the case of a junction between sewers made by a curve in a manhole (see Plate VIII, Fig. 5). This formula, however, does not apply to the foaming or impact created by an angle. A very considerable loss of head may result from this, and consequently sharp bends should be avoided unless it is desired to reduce the velocity.

The obstructions to flow offered by manholes, house-connections, etc., can be almost entirely avoided by careful designing and construction. That due to roughness of the material of construction should also be kept low, but will necessarily be considerable. This obstruction should be allowed for in the formula by modifying the value of c through the different values of n.

Art. 22. Limits of Velocity.

The formula for the quantity of sewage which will flow through a given sewer per second is $Q = Va$, in which a is the area of the stream flowing. It would appear that, given Q, V and a could take any value so long as $Va = Q$. a is, however, limited in its maximum by economic considerations, also sometimes by practical ones (see Art. 23). V also, although, if pure water were the material flowing through the sewers, it might vary from 0 to infinity, is limited within a comparatively narrow range by the character of ordinary sewage.

House-sewage contains some matter which is slightly

heavier than water, also much which is lighter; the former tends to settle in the bottom of a sewer, the latter to collect along the edges of the stream. Ashes, garbage, clothing, and other refuse matter should be kept out of the sewers by laws rigidly enforced, but in spite of all precautions such material will at times reach them. Dirt and sand frequently enter house-sewers through the ventilation-holes in manhole-heads or through defective joints in the sewer. As no system is perfect or perfectly managed, provision should be made for a certain amount of such matter. It is found that if the velocity of a stream be sufficiently great matter suspended in the water will not be deposited, but a retarding of the velocity at any point may cause a formation of deposits there. Experiments have been made to determine the velocities necessary for flowing water to render it capable of transporting matter of various sizes and densities, though usually earth, sand, gravel, and stones have been used. The results obtained by DuBuat are those usually quoted, and are given as being approximately correct for channels of uniform cross-section. The velocities are those sufficient to move the particles along the bottom of the channel and are in feet per second.

TABLE No. 15.

MATERIALS MOVED BY WATER FLOWING AT DIFFERENT VELOCITIES.

Material.	Bottom Velocity.	Mean Velocity.
Pottery-clay	0.3	0.4
Sand, size of anise-seed	0.4	0.5
Gravel, size of peas	0.6	0.8
" " " beans	1.2	1.6
Shingle, about 1 inch in diameter	2.5	3.3
Angular stones, about 1½ inches in diameter	3.5	4.5

Other experiments have given slight variations from these figures, but they are sufficiently accurate for ordinary use. It must be remembered that they apply to loose material only. Where clay or sand has formed a compact deposit in a sewer many times these velocities may be required to move it. Just which of these or similar materials the sewage should be given sufficient velocity to hold suspended is a question. But it has been found in practice that an actual velocity of 1½ feet per second will ordinarily suffice to prevent deposits where house-sewage alone is admitted.

Where storm-water from the streets is admitted to the sewers clay, sand, gravel, leaves, etc., as well as lighter matter are washed through the inlets. The velocities in these sewers should be sufficient to prevent the deposit of such material, which velocity, according to the table given above, would needs be about 3.5 feet per second.

The velocity given for house-sewers—1½ feet per second—is that which should be maintained as a minimum by the ordinary minimum daily flow; that for storm-sewers—3.5 feet per second—is the least which should be attained in time of storms.

The average daily flow in house-sewers may be taken (Art. 14) as ¼ of the maximum to be provided for, and the ordinary minimum as ½ of this. At night-time, when the absolute minimum usually occurs, the sewage is composed of comparatively pure water and a lessened velocity due to a shallower flow will not be particularly detrimental. ⅛ of the maximum volume for which the sewer is designed may therefore be assumed as that for which the velocity should be 1½ feet per second. For reasons to be given (Art. 23) a house-sewer is usually designed to be 50% to 100% larger than required by the assumed volume of sewage, so that the ordinary minimum can be taken as being ¼ to ⅛ of the capacity of the sewer. Reference to Table No. 13 shows that this quantity is carried when the depth of flow in the sewer is .25 to .3 the diameter and when the velocity is .65 to .72 that for a sewer flowing full. It follows from this that the grade of a house-sewer should be such that the velocity when flowing full is at least $\frac{1.5}{.65}$ to $\frac{1.5}{.72}$, or 2.3 to 2.1 feet per second.

In the case of storm-sewers, which carry no house-sewage and are thus dry for a large portion of the time, it may be assumed that in general any storm which will wash any considerable amount of gravel and dirt into them will require at least one third of the capacity of the sewer. Such grades should therefore be given these as will cause a velocity of at least 3.5 feet per second when the sewer is flowing one third full, or 4 feet when flowing full. Smaller showers, which will give less depth of water in the sewer, it may likewise be assumed will contribute only such matter as is transported by less velocities.

It may in some cases be necessary to construct sewers giving somewhat lower velocities than these, but this should be only after careful consideration of the problem. House-sewers should never be designed with grades giving a less velocity than 2 feet per second when flowing full, nor storm-sewers with those giving less than 3 feet.

Where a combined sewer is in question—i.e., one which daily carries house-sewage, but which also has sufficient capacity for and acts as a storm-sewer—the requisite velocity must be obtained for both house- and storm-sewage. But except in very unusual instances a grade which will meet the requirements of house-sewage will more than satisfy the demands of storm-water transportation. For, since the maximum amount of house-sewage per second per acre in a residence district will be about $\frac{80 \times 175}{7.48 \times 86400} = .022$ cubic feet, while the storm-water from such an area may be 3 cubic feet per second, or 140 times as great, if a circular sewer is designed to give a velocity of $1\frac{1}{2}$ feet when it is carrying .007 of its full capacity its velocity when flowing full will be about 9 feet, or more than twice the desired velocity; while with an egg-shaped sewer under the same conditions a velocity of 4.7 feet when flowing full is obtained.

The subject of maximum velocities has received but little attention, probably because the dangers connected with excessive velocities are not so great as those resulting from a too slow rate. Such dangers do exist, however. The more immediate one is that the consequent shallowness of the current which would in many cases result would occasion the deposit of the larger floating solids, which may result in obstinate obstructions in the sewer. In the mains this can be obviated by reducing the size of the sewer to the point where the necessary depth is obtained. But it is usually not in the mains but in the branches that steep grades are possible. To reduce the sewer to such a size as would give any considerable depth to the daily flow on very steep grades would call for a diameter much below that usually adopted as a minimum. An 8-inch sewer whose grade is 0.1 gives a theoretic velocity of 10.04 feet per second when flowing full. To secure a flow in this pipe having an average depth of 4 inches would require the sewage from a population of 6500. In general it may be said that the ordinary depth of flow in any sewer should not be less than 2 inches, nor should it be less than $\frac{1}{2}$ the radius of the invert, since if it is so there is much more danger of deposits forming along the edges and even in the centre of the stream. It will sometimes be impossible to meet this requirement fully, but it should be kept in mind as extremely desirable.

Another objection to too great velocity is the danger of attrition of the sewer-invert by the scouring action of sand, stones, etc., swept rapidly over it. In brick sewers this objection is frequently and successfully met by lining the invert with granite blocks. A $5\frac{1}{2}$-foot two-ring brick sewer in Baltimore, 25 years old, was recently found with its invert in one place cut completely through for a width of 12 to 15 inches and badly worn for a height of 2 feet, and many other places were only a little less damaged. In Omaha's brick sewers the wear, which is usually 18 to 24 inches wide, became 2 to 3 inches deep in 12 years. In both cities ordinary brick was used, but was replaced with stone blocks.

The first objection is the serious one, since the time taken to wear out a sewer-invert must be considerable if good material is used, and replacing it is a matter of expense only. But the forming of deposits in the sewer endangers the health of the community.

It is difficult to set a maximum limit to the velocity allowable, but it may generally be taken as from 8 to 12 feet per second. From 3 to 5 feet per second is probably the most desirable velocity.

Art. 23. Size of Sewers.

If a house-sewer were constructed to exactly meet the theoretical requirements as above outlined it would continually increase in size from the head to the outlet, by a small increment below each house-connection, by a larger one below each tributary branch or lateral; but between the first two connections it should be of sufficient size to carry the sewage of one house only, which would be about $\dfrac{6 \times 175}{7.48 \times 86400}$ = .0016 cubic feet per second, which at a velocity of 2.5 feet per second would call for a pipe of .00064 square feet area, or $\frac{1}{3}$ inch diameter.

This method is not closely followed for the reasons that the data on which are based the calculations of volume of sewage as well as the formulas of flow cannot be exact enough to warrant it; that the estimate of ultimate population may be exceeded; that the per capita water-consumption may increase beyond the maximum assumed, factories or other large contributors of sewage locate at points where they were not expected, or for some other cause the amount of sewage reaching any lateral may be largely exceeded. This excess can be allowed for in a general way only, but it is advisable to design the laterals of a capacity double that calculated, particularly since the cost is not thereby largely increased, and the velocity in a sewer flowing half full is as great as that in one flowing full.

The house-sewer mains need not have so great an excess of size, since they carry the sewage from many laterals, and it is not probable that *all* these will receive double the calculated amounts of sewage. It will probably be sufficient to increase these by 50% of the estimated capacity. The volume of sewage reaching the trunk or outlet sewer can be still more closely calculated, and an increase of 25% may be made as giving it sufficient capacity, although it would probably be better to add 50% here also, the additional cost being slight in most cases.

With this increase the head of each lateral would still be less than $\frac{1}{2}$ inch in diameter. This would be too small to adopt in practice for several reasons: because an individual house will contribute sewage at occasional maximum rates far exceeding 175% of their daily average; because a very small sewer would be too frequently stopped by pieces of paper, cloth, or other legitimate sewage matter; and because it would be too difficult of access for inspection and cleaning. The last two objections could, it is true, be met theoretically by making the house-connection of a size so much smaller that nothing could pass it which would obstruct the sewer. But such construction would be utterly impracticable.

There is no particularly good reason, however, why a house-connection might not be made of 2-inch pipe and the sewer of 3-inch or 4-inch; and systems are in existence and reported working satisfactorily where such sizes are in use. But such construction would generally compel a change in the stock dimensions of all house-plumbing and connected appliances, and give rise to inconveniences more than balancing the saving in cost. A 4-inch house-connection is, however, ample for any building containing less than 50 persons and which contributes only ordinary house-sewage (see Art. 82).

The sewer might, then, where the grade is quite steep, be constructed as a 4-inch pipe from the head to such point as the calculations fix for an increase in size; but it is better to make the minimum diameter 6 or 8 inches, for then there would be less probability that anything passing the house-connection, in which the velocity may be considerable, would obstruct the sewer. It is thought that the weight of evidence tends to show that with 4-inch house-connections 8-inch sewers are obstructed much less frequently than are 6-inch. Among other reasons for this is the fact that a 6-inch stick, chicken-bone, etc., will pass a 4-inch trap, but an 8-inch one will not; and that a 6-inch stick is more apt to become wedged across a 6-inch pipe than across an 8-inch one. Some engineers set the 6-inch, more, probably, the 8-inch pipe, as the minimum to be employed for sewers. In England 9 inches is generally the minimum size.

In the case of storm-sewers the only change of conditions affecting the volume of sewage which is likely to occur is in the imperviousness of the contributing area. If this is taken at the maximum, as for a business district, no allowance need be made. In any case the allowance for change can best be made in the selection of the factor of imperviousness and the sewer built of corresponding capacity. It is probable that no condition of size or character of tributary area will in actual practice call for a **storm-sewer** of a diameter less than 10 or 12 inches. It should, if possible, be of a diameter at least as great as that of the largest opening in the storm-water inlets, to prevent sticks lodging across it.

A circular or egg-shaped sewer is sometimes limited in size by the amount of covering necessary and the distance below the street-surface of its invert, where this is fixed by the elevation of the outlet and the necessary grade from that to the point in question. If the whole sewer at this point be lowered the grade and velocity become less and the size of the sewer must be increased, thus raising the crown. The size can be reduced only by increasing the grade, which means raising the sewer. Under these conditions the sewer can be built as an "inverted siphon" to flow under a head (Art. 38), two or more parallel sewers can be substituted for the one, or the shape can be modified. In adopting the last alternative engineers have devised many forms which can be generally classified as those flattened on the bottom and those flattened at the top.

ART. 24. SHAPE OF SEWERS.

Of all possible shapes of sewers of equal area of cross-section the circular gives the greatest velocity when flowing full or half full and, having the shortest perimeter, contains the least material. Also, being devoid of angles, it offers little opportunity for deposits. For sewers intended to always flow at least half full it is therefore the most desirable shape. This is not true, however, of a combined sewer—that is, one which carries both house-sewage and storm-water—for, as we have seen (Art. 22), the house-sewage may occupy only $\frac{1}{140}$ of the capacity of the sewer and have a velocity only about $\frac{1}{8}$ as great if a circular sewer be used. If the sewer, considered as a storm-sewer, be given a grade adapted to a velocity of 4 feet per second when flowing full or half full the velocity of the house-sewage would be about $\frac{3}{8}$ of a foot per second. If on the other hand the grade be so increased (which is seldom possible) as to give the minimum house-sewage flow a velocity of $1\frac{1}{2}$ feet per second the depth of this flow would be only about .02 of the sewer diameter. Neither of these conditions is permissible in a good sewerage system.

The result of adopting too flat a grade is shown by the illustration (Plate VII, Fig. 8) of obstructions in the old London sewers, which came to be known as "sewers of deposit." These required frequent cleaning, since almost the entire sewage matter was deposited in them, and became very dangerous to the health of the city. The question thus forced upon the attention of engineers was first solved by building in the bottoms of the old sewers channels of much shorter radius of curvature (Plate VII, Fig. 7). These, by increasing R and consequently V, as well as the depth of flow relative to the invert radius, had the same effect upon the flow

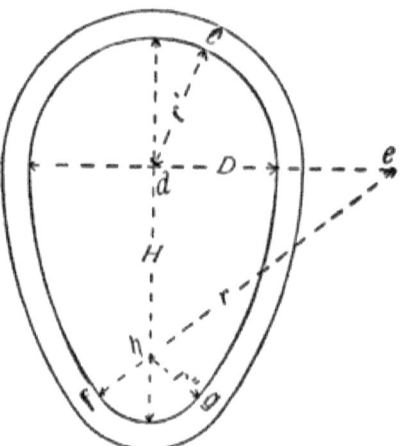

FIG. 2.—EGG-SHAPED SEWER.

as the use of smaller sewers, which they in fact were, and answered the purpose, practically the same design being still employed in Washington, D. C., and other American cities. It will be noticed, however, that there is considerable useless material in this design; also that the bench on either side of the small channel offers opportunity for the deposit of material, which may there putrefy. To meet these objections the egg-shaped sewer was designed and is used extensively for combined, and often for storm-water, sewers. Several proportions have been suggested and used, but that most frequently found in modern American practice is represented here. The diameter of a circular sewer having an equal area is $1.209D$. In this sewer

$$H = 1.5D, \quad dc \text{ or } r' = 0.5D,$$
$$cf \text{ or } r = 1.5D, \quad gh \text{ or } r'' = 0.25D.$$

Reference to Table No. 14 shows that a flow of $\frac{1}{140}$ of the full capacity of this sewer would have a velocity about 0.3 as great as if the sewer flowed full, or 85% greater than the same amount in a circular sewer of equal total capacity; also the depth would be about $0.1D$, or $0.4r''$. If the velocity of the house-sewage in the above be $2\frac{1}{2}$ feet per second (as it should be) that when the sewer were full would be 8 feet or more per second. This form does not, therefore, quite meet the requirements of a combined sewer, intended to carry a run-off of 3 inches from the area drained, as to either depth or velocity of house-sewage. As we shall see later, this requirement applies to lateral combined sewers only, and this design is suitable for most combined-sewer mains, whose maximum flow is only $1\frac{1}{2}$ or 2 inches run-off from the drainage-area. In laterals or other sewers, however, where the proportion of house- to storm-sewage will be too small, or for some other reason sufficient velocity and depth for the house-sewage cannot be thus obtained, the adoption of an egg-shaped sewer with $r'' = \frac{1}{6}D$ or $\frac{1}{8}D$, or a form similar to that shown in Plate VII, Fig. 2, is recommended, the purpose being, whatever the form adopted, to get a satisfactorily high value for R for the house-sewage flow. Whatever the radius of invert the grade must not be less than that which would be required by a circular house-sewer having a radius $= r''$. The radius r'' should be so chosen, also, that the depth of house-sewage will never be less than $\frac{r''}{2}$. A flat bottom should never be used for house or combined sewers unless the sewage will always be sufficient to cover it at least 6 inches deep. Angles in the section are to be avoided as favoring deposits. In storm-sewers it is advisable that the shape be such as to give good velocity to small amounts of storm-water, but the penalty of not following this rule is not so serious as in the case of house-sewers.

CHAPTER V.

FLUSHING AND VENTILATION.

Art. 25. Necessity for Flushing.

It is seen from Table No. 13 that if at any time the flow in a circular sewer becomes less in volume than $\frac{14}{100}$ the full capacity of the sewer the depth becomes less than $\frac{1}{4}$ the diameter and the velocity less than $\frac{2}{3}$ that for a full sewer. If the sewer is small the first condition is apt to cause deposits by the stranding of floating matter on the edges or even in the centre of the stream; if the grade is near the minimum the velocity becomes less than is desirable and deposits result from this cause. But a 6-inch or 8-inch pipe is usually the minimum size employed and is carried up to the last house-connection, from which a quantity of sewage very much less than $\frac{14}{100}$ of the full sewer capacity is received. In fact there will be in a residence district a stretch of at least 400 feet of 6-inch or 700 feet of 8-inch sewer, even at the flattest allowable grade, which would be filled less than $\frac{14}{100}$ of its capacity by a *rate* of 175 gallons per capita, and consequently where deposits are probable. The discharge from any individual house comes usually not in a continuous flow, however, but in spurts of relatively large quantities separated by considerable intervals of time. If we watch such intermittent discharge we will find that when the sewage enters an empty sewer from the house-connection it flows both down the grade and also up it for a short distance. The latter portion at the end of the discharge also flows down grade, but it has probably carried with it and left at the upper limit of its flow matter which remains there to putresce and perhaps form the beginning of an obstruction. Beginning in the sewer at practically nothing (since most of the initial velocity is destroyed by foaming), the velocity of such discharge continually increases, and the depth decreases, with the distance from the point of entry. This frequently causes the stranding below the house-connection of large floating matter which is introduced from such connection, and although successive discharges may move this matter, each one a little further down the sewer, a long cessation of them may give it an opportunity to become fixed in its position. Discharges from connections higher up the grade will tend to prevent these deposits, two or more discharges occasionally coming simultaneously and uniting their volume; and generally the further any connection is from the upper or dead end of a branch the less the danger of its causing such deposits. In a thickly settled district this danger in the case of 6- or 8-inch pipe becomes very small at a point to which there is tributary 1000 to 1500 feet of sewer. If the district is sparsely settled, however, the danger may exist for many times this length.

Any house-sewer, but particularly a lateral, is liable to partial stoppage at times, due to ashes, sand, or other material introduced through house-connections, manholes, or infiltering through the joints or other defective places. Unless the velocity of flow is sufficient to carry this matter along it will form deposits in the sewer-invert which must be in some way removed.

There is another class of deposits, composed of mycelial matter, which forms in most house-sewers. This contracts the area of cross-section and may become the breeding-place of micro-organisms; but emits little odor and is readily detached and carried away by a strong flush of water.

To *prevent* these deposits the only practicable way known is to keep all sewers constantly flowing with a depth at least ½ the radius of the invert, water being introduced for this purpose if necessary, and also to maintain a velocity of at least 1½ feet per second. To *remove* them the methods employed are either to occasionally turn through the sewer streams of water of sufficient quantity and velocity to dislodge and remove the deposits, or to employ shovels, hoes, "pills," scrapers, or similar appliances to be described in Chapter XV.

The method of prevention, if applied near a dead end, where the sewage flow is minimum in quantity, even in the case of a sewer laid at minimum grade, would require about 47,000 gallons per day for each line of 6-inch pipe and 83,000 gallons for each 8-inch line. These quantities it will usually be impracticable to supply; and were it practicable the addition to the sewage of this amount in each of several branches would compel a large increase in the size of the sewer-mains, and greatly increase the cost of treatment in case this method of disposal was employed. There will occasionally be instances, however, where a convenient stream of water can be utilized to advantage in this way.

It sometimes happens that an old sewer-main or other large drainage-channel is at so flat a grade as to be, in part at least, a sewer of deposit. Flushing can be used to advantage in such a case to stir up and remove the matter deposited. A notable instance of this may be found at Milwaukee, Wis., where 40,000 gallons of lake-water per minute are pumped into the Milwaukee River (the flow of which is largely sewage) for flushing it.

In general a sewer in which there is a continuous flow with a depth of at least ½ the radius of the invert and a velocity exceeding 2 feet will need but infrequent cleaning if legitimate sewage only be admitted. If for any reason or at any time these conditions be not fulfilled artificial cleaning will probably need to be resorted to.

ART. 26. METHODS OF FLUSHING.

As stated, there are two general methods of cleaning sewers: flushing, and by the use of some kind of scraper or similar tool. The latter usually calls for no special provisions in the construction and will be treated of in Part III. Flushing, however, is frequently accomplished by appliances built into the system, and the principles involved are other than those controlling hand labor; it is therefore necessary to consider it in designing. Flushing may be done by hand, by automatic appliances, or by use of rain-water.

By the first the sewer can be flushed from any manhole, as well as from flush-tanks; by the second from fixed points only, usually the heads of laterals; by the third the flushing-water enters from roofs through all or many house-connections, or in some instances the inlets are so constructed as to store the rain-water from the street-surfaces or from watercourses and flush with periodic discharges of the same.

The secret of successful flushing lies in compelling a large mass of water to move at considerable speed down the sewer. If the sewer be less than 24 inches or 30 inches in diameter water should as far as possible completely fill it, that deposits may be removed from its entire circumference and also that the effect of the flush may be felt far down the sewer. With the sewer flowing full bore at the upper end the depth of the water will decrease as the flushing-wave progresses down the sewer, until at some point below, at a distance varying with the size and grade of the sewer, with the head of water at the upper end and the volume of sewage flowing, the depth and velocity of the sewage will be but little affected by the flush.

The initial velocity will depend upon the head and upon

the facility offered the water for entering the sewer. There should be a free and open orifice at the entrance end, and if possible the angle between the inside of the sewer and that of the manhole or flush-tank should be rounded. Speed is of as much value in flushing as quantity, and with a given amount of flushing-water the more quickly it can be made to pass through the sewer the better. In most cases little if any benefit would result should a faucet be left continuously running in each house in a city, but $\frac{1}{1000}$ of the same amount of water used in a proper way would be of great benefit to the system.

Although for creating velocity the head in the flush-tank should generally be as great as possible, it must be limited by the amount of internal pressure which the sewer can stand without rupture. A few years ago a brick sewer in Washington, D. C., was, on account of insufficient size, put under such a head of water by the run-off from a cloudburst that its upper half was completely severed from the lower and the sewer destroyed, and a similar result might follow from too great pressure of flushing-water. With a pipe sewer this danger is not so great. A head of 6 or 8 feet at the manhole or flush-tank—which is more than can usually be obtained—should not endanger a pipe sewer. Brick sewers as ordinarily constructed should not be filled to a point more than 5 feet above the invert or, for those more than 5 feet in diameter, higher than the crown. In no case should the water be backed up a sewer-line to such a height as would flood any connected cellars.

The flushing-water should move down and not up the sewer, since the effect of the latter would probably be to sweep the intermediate deposits nearly to the upper limit of the wave and leave them there to dam the flow. The interval which should elapse between flushings will vary under different conditions. In sewers where there is a constant ample flow of water, where stoppages are few and due solely to accident or design of ignorant or malicious persons, flushing need be resorted to only when such stoppages occur. If it is found from experience that stoppages are frequent or that there is a constant depositing of material in the sewers, or if it is foreseen that this will occur from causes mentioned in the previous article, frequent flushings should be provided for.

In the case of a dead end of a house or combined sewer, or one which has but few house-connections made with it, the flushing should be done once in each 24 or at least 48 hours.

Both separate and combined systems have been built and satisfactorily maintained without flushing at any point oftener than two or three times a year. It is probable that this is possible only where there is considerable ground-water entering the sewers at their upper ends, or where the dead ends occur only in thickly populated districts and on grades a little greater than the minimum herein advocated. There is too little definite information on this subject to justify a positive statement as to when, if ever, flushing at dead ends may be profitably omitted. It is advisable so to arrange every house or combined sewer, where the conditions will be those given as favoring deposits, that it can be satisfactorily flushed.

A few experiments have been made on the actual effect of flushing-water in a sewer, chiefly with reference to the velocity and depth of the flushing-water at different distances from the point of entering. Andrew Rosewater found by experiment with a 400-gallon tank at the head of an 8-inch line of sewer discharging 11 gallons per second that at the first manhole, 200 feet below the flush-tank, the water was 6 inches deep and had a velocity of 5.6 feet per second; 200 feet further the depth was 5 inches, velocity 2.8 feet; and 400 feet further the depth was 4 inches, velocity 2 feet—showing the flushing effect to be practically exhausted in 800 feet. Mr. Ogden, in experiments made in Ithaca, N. Y., in 1897, found that with discharges from flush-tanks through 8-inch pipes of from .89 to 1.1 cubic feet per second the flow was reduced to 2 inches at 1123 feet from the flush-tank in two cases where the grades varied from .52% to 1.31%, at about 1000 feet in another where the grades varied from 1.02% to 3.14%, and in another where the grades varied from .80% to .89% the depth was 4 inches at 895 feet from the flush-tank. In the first two the sewer was scoured clean for 529 feet and some effect felt at 819 feet; in the third the sewer was cleaned for 556 feet and the effect slight at 970 feet; in the last the pipe was "disturbed, but not cleaned," at 636 feet, until 600 gallons were discharged, when it was cleaned for more than 636 feet, but less than 900 feet. The other discharges referred to were of 300 gallons each. An interesting series of experiments were conducted and their results plotted by S. H. Adams in England. These appeared to show, as do the above, that 300 gallons is in some cases insufficient to properly flush an 8-inch pipe; also that the effect of such a quantity is felt for about 800 to 1000 feet.*

In flushing by hand the sewer is usually stopped at the down-grade side of a manhole or flush-tank, this is filled to a desired height with water or by allowing the sewage to accumulate in and above it, the gate, plug, or other stopper is removed and the water allowed to enter the sewer under the head due to its height. Where outside water is used for flushing and is limited in quantity another stopper should be placed at the upper orifice in manholes, to prevent a flow up the sewer, and left in until the flushing is over. The stoppers are made of various forms and to act in various ways, and to close the whole or only the lower half or two thirds of the sewer. The water is obtained from different sources and introduced by different methods, a further discussion of which will be given in Part III.

In England the separate system, when first constructed,

* See also Transactions Am. Soc. C. E., vol. XL, pp. 1-30.

was designed to admit to the house-sewers roof-water and drainage from yards, and this method is still followed there to a considerable extent. In the United States the majority of separate systems are not supposed to receive this water. It is argued by advocates of the former practice that the householder should not be required to construct two connections, one for house-sewage and one for rain-water. But the last can be conveniently discharged into the gutter, except in the case of buildings covering a large area, when the cost of the extra drain would be relatively inappreciable.

Another argument for the admission of roof-water is that it is beneficial in flushing the sewer. If it is admitted only at and near the dead ends it will usually be advantageous, but it should not be thought to take the place of all other flushing. The sewers are most likely to need flushing at dry seasons, and this must then be done by hand or otherwise. There is a danger that the presence of these roof-connections will give a false idea that the flushing requirements have been entirely met.

If roof-water is admitted to small sewers throughout their length there is great probability of its gorging the pipes and backing up into connected basements and cellars. In Mount Vernon, N. Y., in 1892 great damage was caused in this way and all roof-drains were at once disconnected; and many similar instances might be cited.

Since the danger is so imminent and the benefits contributed at such uncertain intervals, most American engineers do not advise the admission of roof-water to small sewers.

Sewers are sometimes flushed by connecting their upper ends with convenient streams, or artificial channels filled from such streams, the water being admitted periodically by gates: as at Bern, Wurzburg, Innsbruck, Freiburg, Breslau, Munich, and other cities of Europe; also at Newton, Mass.

Reservoirs fed by streams or springs are used in Munich,

Cologne, Wiesbaden, Frankfurt, Stuttgart, and other cities. At the first-mentioned place large underground reservoirs, one of which is 6 feet 6 inches by 4 feet 7 inches and extends along two blocks, are filled from the Isar River.

Tides are sometimes made use of for this purpose, the water being allowed to rise in the sewer at high tide and being held there by gates until the low tide, when it is released. Ordinarily only the lower reach of the outlet sewer can be thus flushed. A better method in some cases is to hold the water after high tide in a basin from which it is rapidly discharged at low tide into the sewers to be flushed.

As in the case of Milwaukee, already cited, and of Bremen, the flushing-water may be pumped from a lake or river directly to the sewer. This is of course applicable within the limits of economy to very large sewers only, or to a system where a number of dead ends can be reached by a comparatively short line of water pipe.

The water for flushing is sometimes taken from the ocean or other body of salt water; but the salts are thought to decompose the sewage, giving rise to gases and deposits of matter rendered insoluble, and are corroding to any metal-work in the sewers. Hence its use is not advised by most authorities.

ART. 27. APPLIANCES FOR FLUSHING.

Automatic flush-tanks are in use in a large number of separate systems, but are seldom used for flushing combined or storm-water sewers, owing to the enormous quantities of water needed for that purpose. There have been a great number of devices invented for flushing. Most of those at present used in any considerable numbers are siphons in principle, so arranged that a tank in which they are set may fill gradually up to a certain point, when its contents are discharged rapidly into the sewer. The tanks are made to contain at the time of discharge from 150 to 600 or even 1200 gallons for 6- to 10-inch pipe sewers. For larger sewers larger quantities are provided. The smaller quantities are of little use. No tank should discharge less than 250 gallons at a time into a 6-inch pipe, and correspondingly larger amounts into larger sewers. 500 to 800 gallons discharged into an 8-inch pipe once in 24 hours would be more beneficial than half of that amount at each of three or four discharges during the same time. It is probable, however, that in sewers calculated for a velocity exceeding 5 feet per second equal efficiency may be obtained with quantities less than those stated.

The tanks should, of course, be water-tight. They are usually built of brick plastered on both the inside and the out, but might be made of wood or of iron. They should be so built and arranged that the water may have the greatest permissible head above the sewer when discharging. (For details see Art. 47.)

The water may be conveniently admitted to the tank through a half-inch or smaller stop-cock connected with the street-main by a supply-pipe passing through the tank-wall. This cock is continually left sufficiently open to cause the tank to fill and discharge at desired intervals. If the water is inclined to be muddy at times the use of too large a supply-pipe will result in the choking of it by sedimentation. It should be of such a size that the quantity to be used in the tank will pass through it with a velocity of 2 feet per second or more.

The discharge-pipe of the tank should be at least as large as the sewer. It would be better to have it a size or two larger and bell-mouthed at the end, but this is seldom done.

The automatic flushing appliances most in use in the United States are further referred to in Chapter VIII. They are, most of them, covered by patent, and the prices range upward from about $12 for a tank to discharge 150 gallons through a 5-inch pipe.

Where automatic flush-tanks are not used some engineers have built into manholes at dead ends 2-inch to 4-inch pipes connected with adjacent water-mains and provided with gate-valves, as at Mount Vernon, N. Y., and Newton, Mass. This is probably the most convenient method of hand-flushing and the cheapest to operate. The cost at Mount Vernon was about $40 for each 4-inch branch and connection.

There are numerous methods of flushing by hose, by water-tanks, etc., many of which are described in Part III.

In flushing by rain-water no special appliances are ordinarily used, the roofs and sometimes the yards being connected in the ordinary way with the sewer.

Special methods involving pumping, some instances of which have been referred to, need no description, since the details will vary with each case.

ART. 28. NECESSITY FOR VENTILATION.

In every sewer there is a space above the sewage filled with air, and this air, it is evident, will generally be far from pure unless kept in motion and frequently renewed. The odor accompanying all sewage, even when there is no decomposition proceeding in the sewer, is communicated to this air; there will frequently be given off some gases due to putrefaction; and it is possible that malefic germs may escape in vapor from the sewage or from deposits in the sewer, to be carried along by the air-currents. This air probably is seldom motionless. It is influenced by the sewage to move down the sewer; it is warmer in winter and often in summer than the outside air, which condition occasions motion when there is communication between the two; it is driven out of or along the sewer by sudden inflows of sewage from house-connections or branches and sucked in by decrease in the volume of flow; near the outlet the direction and force of the wind affect it, driving it up the sewer or sucking it out; last, and most important, it passes into empty or partly empty house-connections and into proximity to, if not into the air of, connected residences. Herein lies the danger. There is no "sewer-gas" which is deadly to human life, but it is known that air which has been confined in contact with decomposing sewage is charged with "an ever-varying mixture of gases; and of those that are deleterious the more prominent are sulphuretted hydrogen, sulphide of ammonium, and caburetted hydrogen; while ammonia, carbonic acid, and occasionally carbonic oxide derived from leakage of illuminating-gas into sewers are present in more or less large proportions." (W. P. Gerhard, "Sanitary House Inspection.")

The least that can be said of these is that they lessen the vitality and prepare the way for easy conquest by diseases that might otherwise obtain no hold upon the system; they should therefore be excluded from all occupied buildings. The danger due to impure air in dwellings has led the New York Board of Health to conclude that "40% of all deaths are caused by breathing impure air." Playfair asserts that in modern hygiene "nothing is more conclusively shown than the fact that vitiated atmospheres are the most fruitful sources of disease." Death rates have been "reduced in children's hospitals from 50% to 5% by improved ventilation."

While the vitiation referred to in these quotations is not that of sewer-air exclusively, this is included among the causes of it and produces the same effect. Unfortunately the most numerous and fruitful sources of the gases are found, not in the sewer, but in the house-connections or soil-pipes, and consequently not directly under the control of the authorities. The methods necessary to prevent danger from these sources will be considered under the head of House-connections (Art. 82).

ART. 29. METHODS OF VENTILATION.

It is evident that the danger from sewer-air may be avoided, or at least lessened, in two ways: by preventing the creation of gases, and by preventing the sewer-air from reaching human beings in dangerous quantities or under dangerous conditions. No method has yet been found for perfectly accomplishing either of these aims in practice, but both may be partially attained.

Aside from illuminating-gas most of the objectionable gases are given off by putrefaction, and the prevention of this in the sewers is therefore most necessary. This is best accomplished by the removal of all sewage to the outlet before putrefaction can begin; and here is seen the advantage of daily flushing, cleaning the upper laterals of deposits before they reach this dangerous stage. The use of disinfectants in sewage for this purpose is seldom advisable, both on account of the enormous cost and practical difficulties of applying them and because the various and changing characters of sewage in different cities and from hour to hour may introduce such matter as will combine with any given disinfectant to produce deposits and gases fully as injurious as those due to sewage alone. The transporting of germs by sewer-air is probably reduced by reducing putrefaction, although there is very little definitely known on this point, it being uncertain even whether disease-germs are carried by sewer-air at all.

To prevent air from the sewer from entering houses two general methods are in use: placing a barrier in the house-connection, and removing the sewer-air through other outlets. The former is one of the aims of the plumber and is usually attempted by the use of traps. The latter has been aimed at by the use of many ventilating devices, in few or none of which has positive action been successfully obtained. A combination of these two methods gives reasonably good results in most cases, a partial obstruction to the air being placed in the house-connection or its branches in the shape of water-sealed traps, and the power of the air to force its way through these being lessened by ventilation.

If the sewer were a tight conduit with no inlets or outlets except through the house-connections and the main outlet the sewer-air must remain constantly unchanged and stagnant, or must find exit and entrance through these house-connections. The first condition is impossible, for the amount of sewage varies from hour to hour and must displace and in turn be displaced by air driven to and derived from some outside source. In case of a sudden discharge of sewage into such a sewer the air will be driven through the only outlets—the house-connections—unsealing the main traps, and the secondary ones also unless these be amply vented. A strong wind blowing up the sewer from the outlet may produce the same result. In addition to other ventilation of both sewer and soil-pipe it is therefore advisable to thoroughly vent all house-traps.

Attempts have been made to constantly remove the air from sewers by either sucking out the foul air or forcing in fresh; that is, by producing a current through the sewer to a given outlet by either the vacuum or plenum process. Both have proved failures as well as very expensive. In no experimental case has the effect been felt more than 1000 feet from the fans or other apparatus, not only on account of the great amount of air in the sewer-mains and laterals to be moved, but because the traps in the house-connections were unsealed by the pressure and air admitted from or forced into the buildings, according to the system employed.

The Metropolitan Board of Works, London, concluded,

after exhaustive study of the question, "that the method of ventilation adopted in mines, where there are only two openings to be dealt with (an inlet for the air at one end and an outlet for it at the other), is inapplicable to sewers." This characteristic of a sewerage system renders impracticable all methods of ventilation depending upon one or two ventilators to each line of sewers: such as connecting the sewer-end with a chimney, which would afford little more ventilation than an untrapped soil-pipe at the same point or a special ventilating-manhole.

Many expedients for ventilation have been devised and tried—among them connecting the sewers to street-lamps, where a suction is caused and the gas burned by a constant flame; placing in the crown of brick sewers small perforated pipes connected with "uptake-shafts," expected to cause a continuous removal of the gases; leading pipes from the sewer to special flues constructed in houses, within the body of the walls, adjacent to the chimney, or upon the outside of the house and running up above all windows; leaving the main house-drains untrapped and extending them above the roofs; placing flap-doors in the sewers, opening downward for the sewage, but closed to air, which can escape through openings just above such flaps; placing in the street centre at intervals along the sewer manholes or other ventilating-shafts with perforated covers; connecting the sewers by untrapped pipes with street-inlets at the curb line. In connection with these charcoal and other deodorizers are sometimes placed at the air-outlets. (See "General Conclusions, Metropolitan Board of Works," London.)

There seems to be evidence in favor of the conclusion that the greatest danger exists in the house-connections themselves and not in the sewers, although the latter should be prevented from contributing to this danger. Of many analyses of sewer-air made not one to the author's knowledge has shown a greater impurity than that in a crowded city street, whether CO_2, oxygen, or bacteria be taken as the basis of comparison. Equally positive proof goes to show that the average house-connection or the adjacent soil near open joints in the same does give rise to dangerous gases. (It is probable that the upper ends of branch sewers, if not flushed well and often, are open to the same charge.) However, a rush of comparatively pure air from the sewer *forced* through the traps of a foul house-connection is as objectionable as though it itself were polluted, since it forces into the building the impure air existing in such connection. The vents on all traps should hence be of such capacity and so placed as to give full and immediate passage to all the air necessary to prevent forcing or siphoning of traps.

This fact, that the house-connections themselves are fully as foul as, if not more so than, the sewers should be more generally recognized and better provision made for ventilating them. This is reasonably well done by placing a vent-shaft just above the main trap, continuing the soil-pipe above the roof and venting each trap throughout the house. But a still better circulation of air is obtained by omitting the main trap altogether and permitting the air from the sewer to pass through the house-connection unobstructed. The danger of this air passing the traps on house-fixtures is no greater than that of the soil-pipe air doing the same, and in the majority of cases the sewer air is the less dangerous. Such construction is also of great assistance in ventilating the sewer. If only an occasional house-connection be left untrapped, however, the odors from this may be objectionable, the sewer air being but little diluted by the infrequent openings. But the author knows of no city which makes this method compulsory in all connections where it is not perfectly satisfactory. (See also page 344.)

The use of street-lamps as outlets may be advantageous, but the electric light has rendered argument for and against this plan obsolete. The use of hollow electric-light poles has recently been introduced in Columbus, O., with what success it is too early yet (1899) to state. The use of flap-doors in the sewers presupposes a regular flow of air in a fixed direction through the sewer, which investigation has found does not ordinarily exist; this, however, may be advantageous on steep grades, where there is a tendency for the air to rise past intermediate ventilating-points to the highest ones. Ventilation through manholes and other ventilating-shafts most, if not all, engineers recommend, although many do not consider these sufficient.

The use of storm-water inlets for this purpose is much opposed by many, who contend that the sewer-air should not be discharged so near to passers-by upon the sidewalk. In fact this same argument is used by a few against ventilation through manholes in the centre of the street. It is probable that the danger from this cause is very slight, if it exists at all, since it is dependent, not upon the gases, which are enormously diluted upon reaching the outer air, but upon the presence of disease-germs in the exhalations, which is not proven. Moreover, the average catch-basin, even if just cleaned (as this cleaning is ordinarily done), is more offensive than any rightly designed sewer is at all likely to become; and it is extremely doubtful if, in connection with its odors, any contribution of air from the sewer could be detected. For these reasons it seems to the author desirable to connect the sewer with the street-inlets by ventilating-pipes and to place manholes with perforated heads at intervals. Since the latter are apt to be sealed in winter by ice and snow, and in summer by mud, the additional ventilation through the street-inlets would seem to be advisable, particularly if the sewer be not ventilated through the house-drains. A small amount of snow will not ordinarily stop the openings in a manhole-cover, owing to the warm air of the sewer, but a heavy storm or frozen mud may easily do so.

Since the proportion of air in a small sewer to the discharge into the same is much less than in the case of a large combined sewer, and consequently the effect of a given discharge is a greater compression of, and pressure transmitted by, the air in the smaller sewer, the sewers of the separate system need ventilation or safety-vents even more than do those of the combined. In case there are storm-water inlets to which ventilation-pipes from house-sewers may be led this method may be adopted; but ventilation through untrapped house-connections is probably more efficient. This extra ventilation is very often—perhaps in the majority of cases—neglected, but such omission is undoubtedly attended with danger.

For house-sewers, ventilating manhole-heads and untrapped house-drains; for combined sewers, these with the addition of untrapped street-inlets; and for storm-sewers, manholes and inlets—these, with flap-doors on steep grades, seem to the author the best methods so far devised for ventilation; and without ventilation any system will almost surely become a nuisance and a danger. The aim should be to secure by whatever method the greatest possible number and freedom of communications between the sewer and the outer air; and there is little doubt but that when this is realized the sewer air becomes so diluted and the organic matter floating in it so oxidized as to render it less dangerous and objectionable than the air of a crowded church or theatre. When this is not true the sewers are probably in great need of cleaning and flushing.

CHAPTER VI.

COLLECTING THE DATA.

Art. 30. Data Required.

Any plans made before the full and complete data are at hand may be shown by further information to be inadvisable, while their very existence may create a prejudice against the substitution of more efficacious ones. Therefore, although the development of the plans may suggest the desirability of further data the necessity for which was unforeseen, as much as seems necessary in this line and that of surveys should be done preliminary to any designing.

The first necessity will be for a map of the district under consideration. This will usually include the city or town and all land over which it may spread in the future; also all adjacent areas which shed their water into or across the surface of this territory. This map should show all streets, lanes, etc.; all parks or other areas permanently devoted to vegetation; all rivers, creeks, ponds, or other bodies of water—in fact all natural and artificial divisions of the area embraced by the corporate limits. It usually happens that this much can be found already mapped for other purposes; but unless it is known that the measurements from which such map was prepared were accurately taken a sufficient number of check measurements should be made to establish its accuracy or the reverse. On the point of accuracy a question may arise as to how exactly the measurements should be taken. If these should involve an error of no more than .2% they would be sufficiently accurate for the work in hand. For, as sewer grades are ordinarily run from manhole to manhole, and these are about 300 feet apart, an error of .2% would mean that of .6 foot in that distance, which on a grade of .5% (a fairly steep one) would involve an error in grade of .003 foot, which is much less than the least which could be expected in the construction of the sewer.

It will be advisable to obtain also the location of all street-railroads, and of all gas- and water-pipes, their distance from the curb or side lines of the street and the depth of the pipes being noted. Also the location, grade, size, and condition of any existing sewers and appurtenances should be ascertained, by actual inspection if possible.

The data for computing the extent of tributary drainage-areas will ordinarily need to be collected in their entirety, as it is seldom that such information exists in a serviceable form. The topographical surveys which have been made of several of the States, however, may be used to great advantage in this connection. The data desired includes the boundaries of the watersheds whose run-off does not reach a confined channel before entering the limits of the territory to be sewered. (Such water as passes through this territory in the form of streams rather than flowing over the ground does not affect the problem, unless these streams are to be walled in, in which case each one will form a problem by itself.) Also the slope of the ground and the character of the soil as to permeability should be ascertained, the location and extent of rock at or near the surface, of woods and of orchards. Care should be taken to note and locate any slightly worn channels along which storm-water ordinarily flows to the nearest creek or rivulet across territory not yet built up, as these, if they cross into the sewer district, indicate the points at which the storm-water must be intercepted.

Such levels must be taken as are necessary for the plotting of profiles of each street, alley, or any other surface under which a sewer is to run, including a profile across the bed of each stream crossed, with the elevation of high- and low-water marks; also the elevation of the body of water into which either the crude or purified sewage is to be discharged, the elevation during drought and flood as well as the ordinary elevation being ascertained. The depths must be obtained of all cellars whose bottoms are not evidently above the grade of the proposed sewer, unless all sewers are to be placed at a fixed minimum depth, which is to be increased only by the demands of the necessary sewer grades and not by the depth of any cellar or basement (see Art. 37). Also if grades have been adopted for any street, but not yet carried into effect, these as well as the existing surfaces should be obtained.

If a disposal-ground is to be used for filtration or irrigation a careful levelling of its entire surface must be made, and test-pits sunk to ascertain the character of the material to a depth of 5 to 8 feet.

If it is considered desirable to discharge the crude sewage into a given body of fresh or salt water careful search should be made for the point best suited for the outlet; also in case of a river whether the dilution afforded in time of drought will be sufficient to prevent a nuisance. For this purpose the action of currents, tides, and prevailing winds should be investigated. Gaugings of the discharge of streams should be made, and inquiry as to whether and at what points further down the river the water is used for a public supply. It is well also to have analyses made of river-water taken at intervals below the proposed outlet for use in possible suits against the city for nuisance; this whether or not the sewage is to be treated.

The engineer should in person pass through every street in the district to be sewered, noting the character of each, the location of the business and factory districts, the general character of the pavements and yards, and the average size of lot occupied by each residence. He must also ascertain as nearly as may be the present population and its past rate of increase; the probable direction and extent of the future growth of the business part of the city, as well as of the city as a whole. He should obtain the figures, if they exist, of water-consumption in this and neighboring cities; also all possible data concerning the rainfall.

A considerable amount of other information will in many instances be desirable, called for by the peculiarities of each case. Many items, such as cost of materials and labor (for use in the estimate), will suggest themselves as they are needed.

ART. 31. SURVEYING AND PLOTTING.

Since extreme accuracy is not necessary in the transit survey, the use of the ordinary stadia methods will be found advantageous for either check or original surveys. Stadia-hairs in the level, for use in running street-profiles, will be found to expedite this work, and will permit reducing the number in the level party to two. The adjustment of the stadia-hairs should be frequently checked.

The tributary drainage-areas will not need to be surveyed in great detail. If the natural features are boldly accentuated it may be sufficient to locate by a transit-line the limiting summits and ridges, both main ridges and spurs. If the country is gently rolling or generally flat contour surveys should be made of the whole drainage-area, or at least of any portion of it the disposal of whose run-off may offer difficulties.

Of such undeveloped areas as may be reached by the city in its future growth and which will be embraced in drainage-areas for which sewers are to be at once designed accurate contour surveys should be made, contours being located from 1 to 25 feet apart vertically, according to the nature of the country. They should be sufficiently close to show the configuration of the ground in considerable detail, but not so close that the contour-lines will obscure all else upon the map.

Most cities and towns of any size have the street grades established and recorded with their profiles. An extensive experience in attempts at the adaptation of such information to the requirements of sewer-designing has demonstrated that in nine cases out of ten it is waste of time to attempt to use these records and profiles. For the levels have usually been taken by a succession of surveyors of varying degrees of efficiency; occasionally also the grades have been altered on the ground, but not upon the profile; and the time employed in discovering and rectifying errors and omissions would generally have sufficed for taking entirely new levels.

The levels of the street-surfaces taken for the profile need be to tenths of a foot only, but the bench-marks and back- and fore-sights should be to thousandths. Readings should be taken along each proposed sewer-line not more than 100 feet apart, at every pronounced change of grade and at street intersections; the elevation of rails where the line crosses a railway, and at stream-crossings the profile of the bottom and the water-surface, should be obtained.

A convenient scale for a map of a village or borough is 200 feet to 1 inch, but if its size is such that this scale would necessitate the use of paper more than 3 feet wide it may be better to use a scale of 250 or 300 feet to 1 inch. It is inadvisable to use a smaller scale than this, and if the resulting map is still too large for the paper it may be necessary to spread it over two or more sheets. In such a case it will be found convenient, where conditions permit of it, to so arrange the sheets that each drainage-area shall appear upon one sheet only. Upon this map should be shown the location of the proposed sewers and all appurtenances, these being usually in red.

A convenient scale for the profiles is 25 feet to 1 inch horizontal and 5 feet to 1 inch vertical. These should show the sewer-line at its proper grade, the depth of all unusually low cellars, the location of all manholes and other appurtenances. A plan of the street is usually placed under the profile, showing the location therein of the sewer-line and all appurtenances.

For ascertaining the best location for an outlet into tidal waters the use of floats is desirable, since thus can be learned the ordinary periodic movements of the water into which the sewage is to be discharged, and hence the possibility of the creation of a nuisance thereby. These floats should expose as little surface to the wind as possible. A pine rod or tin tube, weighted at the bottom and with a numbered flag fastened to the top, is usually employed. They should be started at different stages of the tide from each point which is being considered as a possible outlet. Account should be kept of and allowance made for winds during the times the floats are in the water. Each float should be numbered and a record kept showing the time and place at which it was put into the water, the state of the tide, wind, etc. By means of one or more boats they should be so traced that the path of each can be plotted upon a map until it strands or passes beyond the point where sewage can create a nuisance. It may at times be necessary to follow a set of floats night and day for three or four days; seldom longer than this, for if they have not in that time passed to a considerable distance from the starting-point such point is not suitable for an outlet.

The quantity of water flowing in a given stream and the resulting dilution can be ascertained by the use of floats or a current-meter, the cross-section of the stream being first obtained. In some cases this flow can be obtained from government or State records of gaugings. If possible a gauging of the stream during a drought should be obtained, since it is even more important that there be the necessary dilution at such a time than when the river is high.

It is sometimes desirable to sink test-pits or bore at intervals along the line of each proposed sewer to ascertain the character of the material to be excavated. This is unnecessary where cellars or other excavations along the street-line have been sunk to practically the depth of the sewer, and when neither rock nor quicksand is anticipated it is seldom of a service commensurate with the cost. In sounding for rock several methods have been used. An iron rod, upset and pointed at one end, may be driven to a depth of 10 or 12 feet through most soils, and may be raised again by a handle, as shown in Fig. 3, which can if necessary be fastened to a lever, a stout wooden horse being used as a fulcrum. It is possible to reach still further by replacing the first heavy rod by a thinner and longer one driven in the same way.

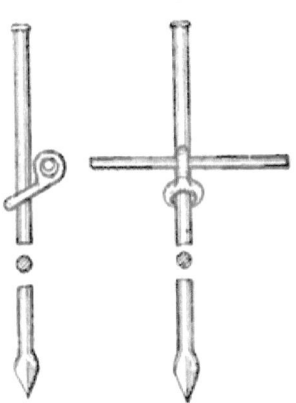

FIG. 3.—SOUNDING-ROD.

When there are not many boulders or gravel-stones in the soil an iron pipe about 1 inch in diameter may be connected by hose with a fire-hydrant and sunk into the ground by the "jet process" to a considerable depth. By connecting the hose to the side of a T screwed to the end of the pipe and capping the top of this the pipe can in most cases be driven by hammer past any small stones or other hard obstacles.

A modified post-hole auger can be used for the same purpose, with the advantage that by it samples of the soils passed through may be obtained.

The only certain method of detecting the presence of quicksand is by sinking a test-pit, though the absence of sand from the materials removed by other methods would of course be proof of its absence. The washings from a jet-pipe may be caught and from the sediment some idea be had of the materials encountered, though not of their consistency.

The presence of ground-water in any quantity is fully as important a matter in designing as the presence of rock, and should be thoroughly investigated. Ground-water is frequently found in porous soils just at the base of a hill. It is usually found in gravelly soils, near hills or mountain streams whose waters percolate into the porous ground. Usually (although there are exceptions) but little water reaches the soil from rivers, whose beds are in most cases impervious. The presence and amount of ground-water can be known only through excavations, which should be made with that aim in view if none exist made previously for other purposes. In many cases a sufficient number of wells and cesspools will have been dug to give a general idea of the depth and amount of ground-water to be encountered.

CHAPTER VII.

THE DESIGN.

No general directions for designing a sewerage system can be given which will cover all the conditions met with in every case. But upon the principles stated may be based any special designs, care being taken to violate none of the requirements of sanitary sewerage. In many cases no emergency will arise out of the ordinary. To such the methods herein outlined apply, but even in the use of these skill and judgment must be employed, and it may frequently be necessary, as it is always desirable, to call upon the services of an experienced consulting engineer for a decision as to some of the vital principles involved in the design; such, for instance, as the system to be employed and the method of disposal. But many small cities and towns cannot afford this expense—or think they cannot—and the city engineer must rely wholly upon himself for the design as a whole and in detail. It is hoped that the principles already stated and the methods following may be of service to him.

ART. 32. GENERAL PRINCIPLES.

The first matter to be decided upon in preparing the design is, How much and what kind of sewage must be provided for? the second, What disposal shall be made of it? the third, What system—separate, combined, or compound—shall be employed?

It is assumed that all urban districts require house-sewerage. Local circumstances, financial, topographical, and geographical, will usually decide whether or not storm-water also shall be removed by the sewers. In small cities there are usually a few places the removal of storm-water from which is almost imperative. These places must be ascertained, the area draining to them measured on the contour-map, and an estimate made of the run-off based upon the principles given in Articles 18–20.

In towns or districts which are closely built up the storm-water should not flow in the gutters more than two blocks—or say 700 feet—before finding a sewer-inlet or some natural stream or channel into which it can discharge. In residence or suburban districts the same rule applies when the streets have impervious pavements and the yards are small. As the pavements become more pervious and the houses more scattered this distance can be considerably increased and the extent of the storm-sewer system proportionately reduced. The judgment as to how many localities (from a lack of watercourses or other reasons) need storm-sewers must be balanced against the funds available for such sewers. If possible, however, the storm-sewers should cover as wide a territory as the house-sewerage system.

In most small cities natural watercourses are retained to carry away the run-off, and the service rendered by these may be made adequate—if it is not already so—by enlarging, straightening, and walling them. (If the money necessary for substituting a storm-sewer for such a drain is available this should of course be done.) The residents along such a watercourse should be prohibited from depositing any excreta, garbage, or other refuse therein; and if this is enforced and the stream so enlarged as to prevent overflowing it will become a good substitute for a storm-sewer, and much less objectionable than such small streams ordinarily are to the occupants of the property it traverses. For the amount of water to be provided for from given areas see Arts. 17–19.

A short summary of some of the principles previously stated may be given here to advantage, with applications of the same.

The amount of house-sewage depends, first, upon the population to be provided for. This must be the population some years in the future; some say 30, some 50 years. The first seems preferable in most cases, since the larger sewers called for by the second will be less suited to the needs of the present, deposits dangerous to health more probable, and consequently cost of maintenance greater; also in most cases the difference in cost at compound interest for 30 years would amount to sufficient at the end of that time to build a system adequate for the increased needs. Moreover, the growth cannot be predicted with any great accuracy 30, and still less 50, years ahead. From the estimate of Baltimore's growth made by the Sewerage Commission it is calculated that to provide for a population for 30 years ahead would call for sewermains of twice the capacity at present required; while if that for 50 years ahead were adopted as the number to be provided for the mains would need to be more than three times such capacity.

For making this prediction it is customary to plot all known past populations, each year and its corresponding population being made coördinates of as many points. A curve is passed as nearly through these points as possible, and with the same law of curvature is continued ahead far enough to cover the time required. It is evident that such a curve should not return on itself horizontally, but must approach an asymptote whose direction the judgment must decide; or the curve may in reality even reverse. This method is but a "scientific guess," but there seems to be no better one. As a general rule the smaller the city or town the greater the probability of sudden and great unforeseen changes in the rate of growth.

The estimate of per capita water-consumption is similarly difficult. There is no necessity for this exceeding 50 or 60 gallons daily, and yet it may reach 200 or even 300 for anything which we know to the contrary. Since it can be confined well within the 100 mark by the use of meters and thorough inspection, it seems wasteful of capacity and capital to provide for more. The probability is that the near future will see the consumption almost universally reduced below this limit.

The population decided upon times the per capita water-consumption and plus the leakage may be taken as the amount of sewage to be provided for.

The character of the sewage, involving the proportionate amount of house-wastes and diluting-water, the character of the water supplied, the presence of acids or other manufacturing wastes, will have a bearing upon the method of disposal.

In deciding upon the disposal to be adopted, if that by dilution is practicable the laws of the State should be investigated to determine its legality; the direction and velocity of tides and currents should be known to be such as to remove the sewage continuously from rather than toward all shores or other places where it may be deposited and create a nuisance; the number of gallons of unpolluted water passing the outlet each day should be equivalent to at least 1500 times the population; the velocity of the water past the outlet must be sufficient to prevent the deposit of sewage matter at or near said outlet. The effect of the discharge upon bathing-beaches, upon fish, oysters, or other food matter, upon the water-supply of towns below, or upon manufacturing interests—these must all be studied, both on their scientific and commercial sides.

If from these investigations dilution is found inadvisable the method of treatment best adapted to the circumstances must be sought. Search should be made for a spot or spots which are low and flat, but not boggy, whose soil is pervious and whose value is low (although land which possesses none of these qualities can be used for sewage disposal), and whose extent is sufficient for years to come. If the sewage must be thoroughly purified filtration or irrigation must be used, alone or in connection with precipitation or septic tanks. Chemical precipitation may be employed alone where a removal of 50% to 65% of the impurities will be sufficient. (See the works on sewage purification referred to in Chapter II, which should be diligently studied before deciding upon any scheme of treatment.)

It will usually be well to make preliminary plans based upon each of two or three methods of disposal and compare them from both sanitary and financial points of view.

A decision as to the system to be employed should ordinarily rest largely upon the decisions of the two previous points. If treatment of the sewage is necessary or will probably become so in the course of 20 or 30 years, or if the house-sewage is to be discharged at some distance from the centre of the city, the separate or compound system will usually be advisable.

If there are a number of convenient points along a water front at each of which house-sewage can be discharged without nuisance the combined system may be the cheapest and most desirable. If there already exist large sewers discharging at various points where the discharge of house-sewage creates a nuisance, or of a character not adapted to carrying house-sewage (because of flat bottoms or rough interior), the separate system will usually be advisable, the old sewers being used in the storm-sewer system. If such large sewers are adapted in interior surface and form to carrying house-sewage, however, they may be retained for this purpose, but an intercepting sewer built to receive from them the dry-weather flow and convey it to a suitable outlet, the storm-water discharging through the previous outlets.

In all these matters, however, engineering experience and judgment, and not fixed rules, should be the basis of decision.

The general rule in sewerage, as in other engineering work, is: obtain the best results and at the least cost. Certainty of attaining this will frequently require the preparation and comparison of alternative plans, both of the system as a whole and of its separate parts.

ART. 33. SUBDIVISION INTO DISTRICTS.

For the purpose of sewerage-designing the territory under consideration is ordinarily divided into two sets of districts, one based upon the density of population, the other upon the slope of the ground-surface.

The former division should take as a basis the probable density of population per acre of different sections at some time—say 30 years—in the future, since the system must serve the population at that time as well as in the present. It will be convenient to base the division upon population per acre of 20, 30, and other factors of 10, 20 being the minimum assumed for habitable districts in most cities. The maximum may run up to 150 or more per acre. As this division is for the purpose of design only and is not usually shown upon the finished map, it may be designated by bounding-lines or by tints upon a working map. (It will be well to have several copies—white or blue prints will do—of the city map as working maps.) Having made the above subdivision, the total population of each area, calculated from the assumed density of population, should be ascertained, and the sum of all these compared with the future total population as estimated by use of the curve (Art. 32). It may exceed this by a small amount—say 10%—to allow for incorrect apportioning of densities. If it does not at least equal it changes in the extent of the different areas should be made sufficient to give this total and at such points as the engineer's best judgment dictates.

The second subdivision is that into drainage districts. For this purpose a carefully prepared contour-map of the city or area to be sewered is necessary. Each district is to contain all the territory draining into one main sewer, together with that main down to its outlet or junction with the intercepting or outlet sewer. Under some plans of sewer assessments this subdivision is necessary for other than engineering purposes. For house-sewers it can usually be best made after the designing of the sewers is completed. For storm-sewers, however, it should be made after the lines are located, but before the sizes are determined upon, to facilitate calculation of the latter.

ART. 34. LOCATING THE SEWER-LINES.

Unless this location is already occupied by gas- or water-pipes or a street-railroad house and combined sewers are in most cases located in the centres of streets or alleys, the cost to the householders on each side for house-connections being thus made equal. In some cities the sewers are located under the sidewalks, there being a line on each side of the street. This plan, which is used at Washington, D. C., quite extensively, is usually adopted in the case of wide streets, since there the cost of the extra line is less than that of the additional lengths of house-connections required by a single sewer. From a financial standpoint the double line is cheaper when the cost of a minimum-sized sewer (6- or 8-inch) of a length equal to the average house-lot frontage is less than the cost of a house-connection of a length equal to the distance between the two sewer-lines. Another advantage of side sewers is that the street-paving need not be torn up in making house-connections. A serious disadvantage is that the distance from the upper end of each line to the point where the sewage flow is self-cleansing in volume and velocity will be double that when but a single line is laid. Also the roots of shade-trees are apt to cause serious trouble by entering the pipe-joints. Probably the best method of avoiding both these last objections and that of the continual tearing up of the street-pavement is to lay the sewer in the street centre and at the same time carry each house-connection to the curb.

Where a city has alleys intermediate between the streets it may sometimes be advisable to carry the sewers through these rather than through the streets, the principal argument for this being that less valuable paving is destroyed and less obstruction caused to traffic by the work of construction. On the other hand the house-connection will be longer, and both the cost increased and the grade in such connection decreased, if the distance from the house to the street centre is less than that to the alley centre, as is generally the case. Moreover, the paving in an alley *should* be equally as good as that in a street, and the unevenness consequent on sewer construction is exceedingly apt to contribute to the disease-breeding slovenliness in what is often at its best an elongated Gehenna. Again, in a narrow alley the space available for piling the excavated dirt is so contracted that the cost of construction is frequently increased by a very appreciable amount on this account. On a side hill, however, it may often be advisable or even necessary to locate sewers in the alleys for the drainage of houses on the lower sides of streets above.

Sewers should be laid in continuous straight lines, as far as possible.

No turn greater than a right angle should be made at any one point by any sewer less than 24 inches in diameter, and any turn whatever made by such a sewer should be in a manhole, by means of a curved channel. For sewers larger than 12 or 15 inches it is advisable to use two manholes in making

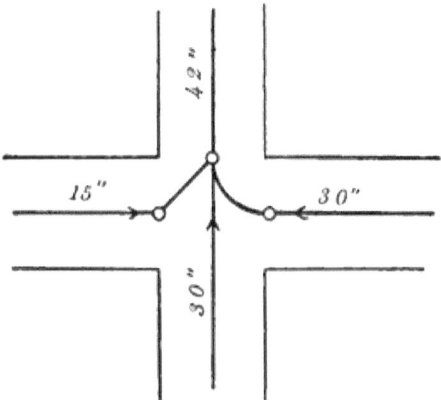

Fig. 4.—Alignment of Sewer-junctions.

a bend greater than 45° (see Fig. 4). Brick sewers more than 24 inches in diameter may be laid on curves, since they can be entered for inspection or cleaning.

Each lateral sewer should take the most direct course to its main, each main the most direct course to its outlet, and the number of mains should be as few as possible. This serves both economy and sanitary efficiency.

The dead ends should be made as few as possible, even at some expense of additional excavation, but not by reducing mean velocities below 2.5 feet per second; nor is it ordinarily serviceable to unite the upper ends of sewers flowing in opposite directions.

House-sewers should be carried within reach, as regards both horizontal distance and grade, of every lot in the sewered district.

Storm-sewers should have as few branches as can be

made to reach all the street-inlets, to better insure which such inlets should be located previous to the location of the sewer-lines.

It is generally advisable to avoid crossing private property where possible, since legal complications and delays might result from such crossing. This will frequently be impossible, however, particularly near outlets.

The sewer-lines can usually be laid out directly upon a contoured working map, an approximate rough estimate of the necessary size and consequent minimum slope of each sewer being made, that deep or shallow cutting may be avoided. The direction of flow should be indicated by arrows.

In the separate system the storm-water sewers should usually be placed on one side of the street centres, the house-sewers being placed in the centres. The two should never be placed one above the other in the same trench unless in contact with each other or connected by masonry.

Art. 35. Volume of House-sewage.

Since the grade of a sewer is limited by its size, and the size is determined by the grade and consequent velocity, but to even a greater extent by the maximum volume of sewage to be carried, this last must be determined before either the limiting grade or size can be decided upon. If the maximum rate of water-consumption be taken at 175 gallons per day per capita the maximum volume per second to be carried by a sewer (in cubic feet) is $\dfrac{175DA}{7.48 \times 86400}$, in which $D =$ density of population and $A =$ the area in acres.

Beginning at the summit of each lateral, it is clear that it is unnecessary to calculate the capacity required for any section of sewer until the point is reached where the volume of sewage to be carried exceeds the capacity of the smallest sewer used at the given grade. For an 8-inch pipe flowing half full with an average velocity of 2.5 feet per second this volume is about $\left(\frac{3.1416 \times 16 \times 2.5}{2 \times 144} = \right) 0.4363$ cubic feet per second, which would be contributed as a maximum flow by a population of $\left(\frac{0.4363 \times 7.48 \times 86400}{175} = \right) 1611$, or about 40 acres having a density of population of 40.

At the point where the sewage from the tributary population exceeds the capacity of the sewer its size must be enlarged to the next market size of pipe or the next size of brick sewer convenient for construction.

The allowance for leakage into the sewer of ground-water, which should be a small proportion of the sewage proper, may be added at intervals, according to the engineer's judgment, based on such data as he is able to obtain.

In calculating these volumes it is advisable to begin with the furthermost lateral sewer first; where this joins another the contributions of both are to be added to determine the flow below that point, and in tracing down this line as each branch is encountered its contribution must be calculated and added. Decision having been made, after a study of the topographical map, as to the line of sewer into which each section of undeveloped territory will drain when sewered, the sewage which this area will ultimately contribute should be placed at the heads of the volumes of flow in this line.

An excellent method of making these calculations is shown on page 122. The sewerage-map Plate No. III was used for this table.

In this case it is seen that the capacity of an 8-inch pipe at the minimum grade was reached at the junction of the Newcastle and Budd Street sewers, but the line down Budd has 1 : 50 as its grade, and no increase of size is yet necessary.

CALCULATION OF SEWAGE QUANTITIES AND SEWER SIZES.

Street.	From	To	Area Acres.	Density.	Population.	Sewage Gallons per Day.	Total Sewage.	Grade.	Size.
Prospect	Newcastle	Walnut	10.4	20	208	36400	......	1 : 30	8 in.
								1 : 11	"
Walnut	Prindle	Prospect	1.7	20	34	5950	......	1 : 20	"
Walnut	Prospect	Liberty	1.9	20	38	6750	......	1 : 20	"
Liberty (extended)	Newcastle	Walnut	12.7	20	254	44450	......	......	"
Walnut	Liberty	Budd	8.2	20	164	28700	122250	1 : 300	"
......	......	......	......	......					
Newcastle	Prospect	Liberty	1.5	20	30	5250	......	1 : 300	"
Undeveloped territory tributary to Newcastle			27.3	20	546	95550	......		
Newcastle	Liberty	Budd	7.5	20	150	26250	......	1 : 300	8 in.
Undeveloped territory tributary to Budd			44.0	20	880	154000	281050	......	
......	......	......	......	......					
Budd	Newcastle	Walnut	11.0	20	220	38500	319550	1 : 50	8 in.
Budd	Walnut	River	3.0	20	60	10500	452300	....	10 "
Budd	Walnut	River			Ground-water	.....	......	....	.....

At the junction of Budd and Walnut the sewage amounts to 441,800 gallons, or 40.9 cubic feet per minute, and the sewer from there to the river must have the minimum grade allowable. The size must therefore be increased, and as the next market size, 10-inch, has a capacity at that grade when two-thirds full of about 590,000 gallons, it is therefore sufficiently large for the rest of the line, including sewage contributed along its length and ground-water. No ground-water was anticipated on the hill side, but it was considered probable that on Budd below Walnut this would leak into the sewer at the rate of two gallons per day per foot of sewer (see Art. 46).

ART. 36. VOLUME OF STORM-SEWAGE.

The principles stated in Articles 16-20 will be used as a basis in determining the amount of storm-water to be provided for. Decision should first be made as to whether this shall include run-off from storms of the first, second, or third class. Then the past rates of fall of such storms should be ascertained. If the records of such rates extending over a series of years are not obtainable use may be made of the rainfall data given in Art. 17. Plate No. IV shows rainfall-curves for average maximum rains of the second class, from

THE DESIGN.

Plate III.

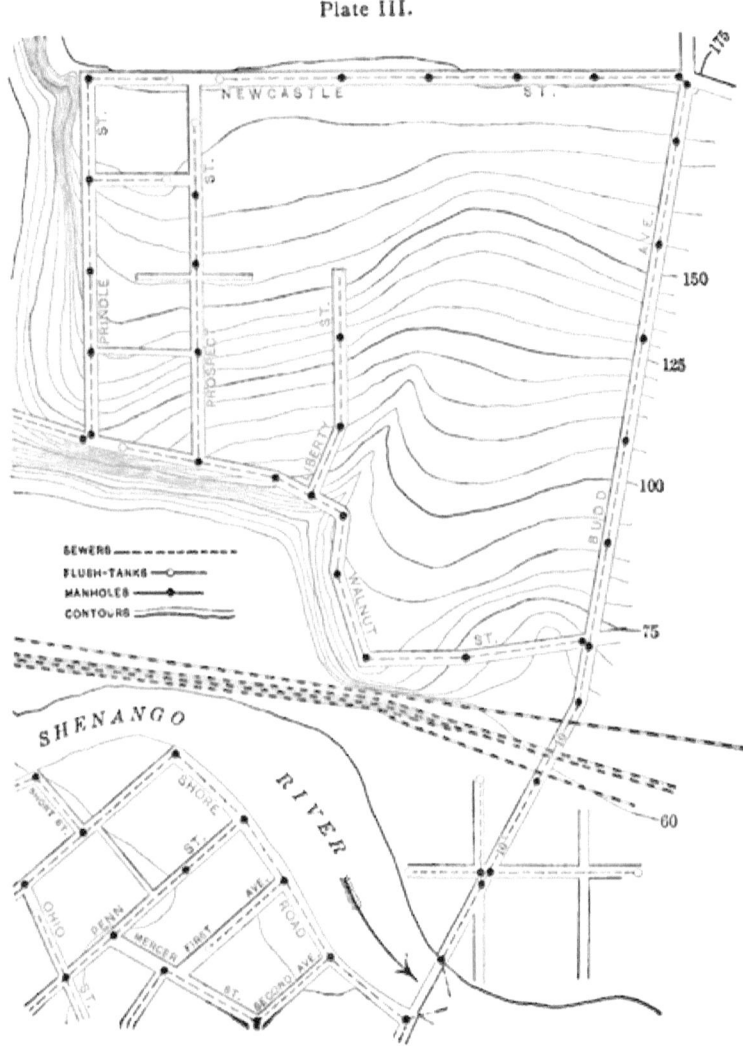

which may be taken the amount of rain to be expected during any given period of time in the localities named. If sufficient rainfall data for the place in question are available a similar curve for that place plotted from these data will be found serviceable. If these data have not been kept by the city it is probable that the rates for a neighboring city can be obtained from the Weather Bureau at Washington, which now has self-registering gauges in over fifty cities of the United States.

Next to be determined is the character of surface of the streets and included areas in each section drained; that is, the amount of impervious surface. The safest course would be to assume that every street-surface is, or will be made, wholly impervious; that the space covered by each building will also be impervious; and that in residence districts the remaining areas will be 30% to 80% impervious at the time of heavy downpours, since rainfall records show that at least 25% of these are preceded by one or more hours of rainfall, which increase the natural imperviousness. These figures will be used in illustrative calculations in this work; but the judgment of the engineer, based on local conditions, may well dictate others, differing for each case considered. For instance, a closely built-up business district having paved yards and courts may be assumed as all wholly impervious.

These points having been decided, the inlets should be located on a contour-map. Also it will be well to state in figures on each city block its area and percentage of imperviousness (see Plate V).

The percentage of imperviousness may be calculated thus:

Let $l =$ the average length of a city block·

$b =$ " " breadth " " "

$f =$ " " number of front feet to a building-lot;

$d =$ " " depth of a building-lot;

THE DESIGN.

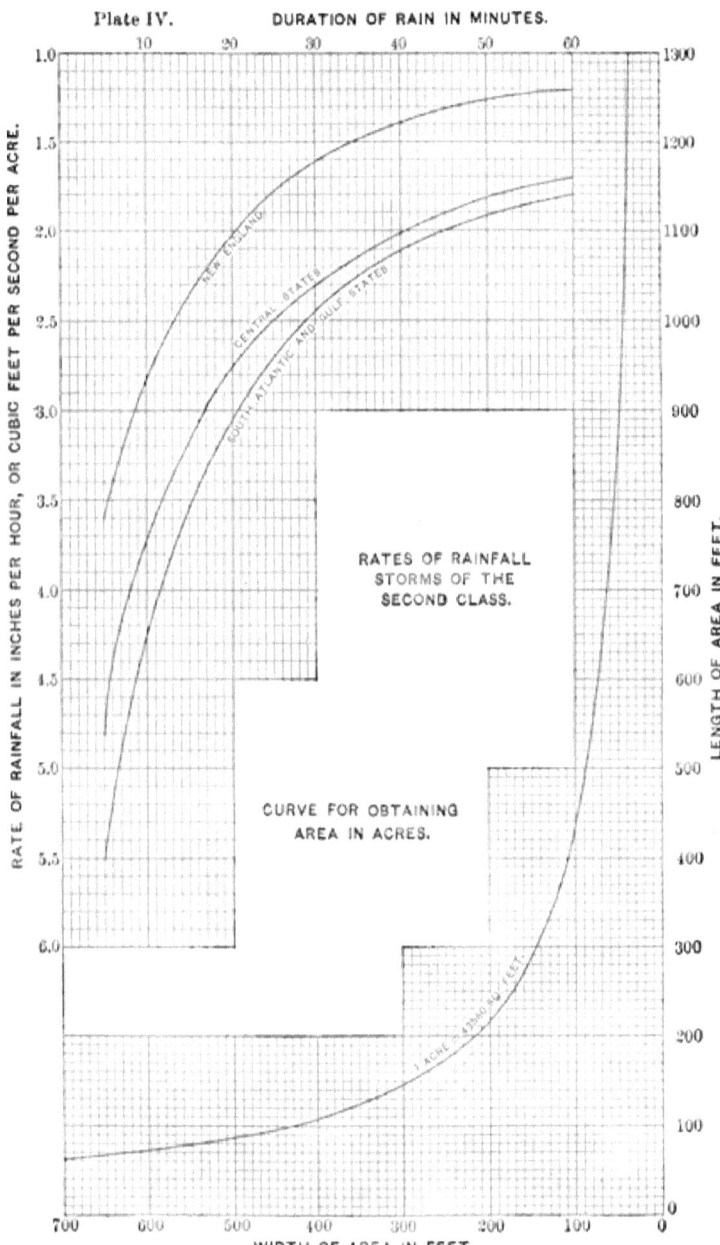

Plate IV.

$a =$ the average area covered by a building;
$w =$ " " width of street;
$i =$ " " percentage of imperviousness of yards, courts, etc., expressed as a decimal;
$I =$ " percentage of imperviousness of the entire area, expressed as a decimal.

Then

$$I = \frac{\frac{alb}{fd} + w(l + b + w) + i\left(lb - \frac{alb}{fd}\right)}{lb + w(l + b + w)}$$
$$= \frac{lb(a + ifd - ia) + wfd(l + b + w)}{fd[lb + w(l + b + w)]}.$$

As an example, let $l = 450$, $b = 250$, $f = 50$, $d = 125$, $a = 1200$ sq. ft., $w = 66$, $i = .60$; then $I = .777$, or say .78.

When most of the above factors must be estimated by judgment only, as for areas not yet opened up or fully developed, it may be as well to estimate I at once.

By comparing this formula with that on page 36 we see that

$$I = P\frac{lb(a + ifd - ia) + wfd(l + b + w)}{43560\,lbo},$$

which formula can be used when P has already been calculated. The relation between I and P will, it is evident, vary in different cities and also in different parts of the same city.

The map having been thus prepared, with the a and I on each block, the uppermost corner of the drainage-area furthest from the outlet may be taken as a starting-point. If there are beyond this any areas not included in the sewered districts, but the run-off from which flows into such districts, this run-off must be estimated and provided for. For this purpose the formula $Q = AIR$ may be used, A being the total area, I the coefficient of imperviousness, and R the maximum rate of rainfall (of the class to be provided for) for that length of time which will elapse while the run-off from the furthest point of the drainage-area is reaching the sewer. This time is an uncertain quantity and will to a certain extent vary with R. Some engineers assume a velocity of about 2 feet per second over the surface. The formula $v = 1000I\sqrt{S}$ is offered as an empirical one for calculating the velocity of run-off over the surface in feet per minute, S being the sine of the slope. While I does not directly affect this velocity, it is observed that the most impervious surfaces usually offer the least obstruction to the flow of water, and *vice versa*. The time t for which R is assumed is obtained by dividing v into L, the length of the furthest corner of the drainage-area from the sewer.

The same method is also applied to determining the time of run-off from each smaller area to its inlet, L in such cases being taken as the distance by gutter of the furthest point from its inlet.

The amount of run-off to each point of interception thus found must be provided for by inlets of sufficient size and number (see Art. 41) and by ample sewer capacity. The following tabulation of a calculation by the above method for the district shown in Plate V is given as an illustration. a is the size of each sub-area, I its imperviousness; AI is in each case the sum of all the preceding aI's. s is the surface-slope of the sub-area, l is the greatest distance traversed by the run-off in crossing each sub-area, t is the time occupied by the run-off in travelling the distance l, r is the rate of rainfall for the time t, $q = aIr$. S is the slope of the sewer removing the run-off from the point in question, L its length to the point next considered (usually the next inlet or sewer-junction), T is the time occupied by the run-off in flowing from the extreme limit of the entire drainage-area A over the surface and through the sewers to the point under consideration. R is the rate of rainfall for the time T, Q is the total

128 SEWERAGE.

Plate V.

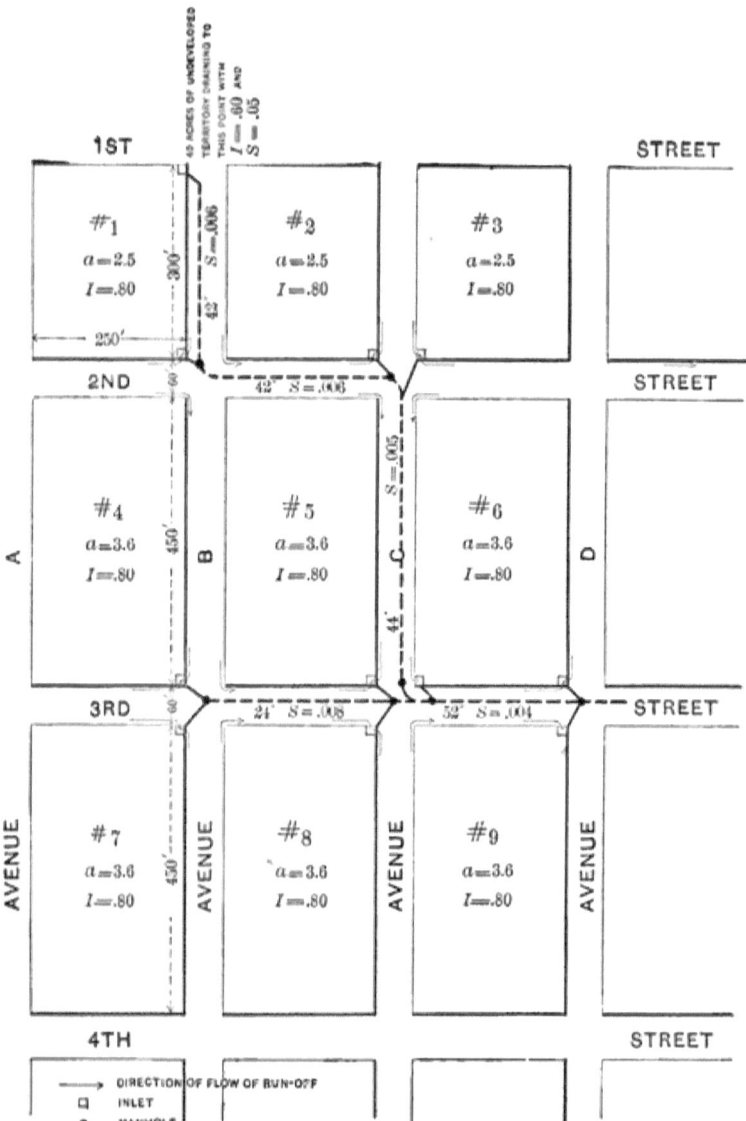

THE DESIGN.

Location of Area	a	I	aI	AI	r	i	t	q	s	T	R*	C	Size of Sewer	Velocity of Flow, Feet per Minute	L	Location of Sewer
Undeveloped territory 40		.60	24	24	.05	3600	16.1	2.3	55.2	.006	16.1	2.3	55.2	42"	420	360	Ave. B, 1st St. to 2d St.
No. 1	2.5	.80	2.0	26	.01	625	6.2	3.4	6.8	.006	16.9	2.2	57.2	42"	420	310	2d St., Ave. B to Ave. C
2	2.5	.80	2.0	28	.01	615	6.1	3.5	7.0	.005	17.7	2.2	66.0	44"	370	510	Ave. C, 2d St. to 3d St.
3	2.5	.80	2.0	30	.01	625	6.2	3.4	6.8								
4	3.6	.80	2.88	32.88	.013	830	3.3	9.5								
7	3.6	.80	2.88	35.76	.007	830	9.9	2.9	8.3	.008	9.9	2.0	16.6	24"	390	830	3d St., Ave. B to Ave. C
5	3.6	.80	2.88	38.64	.005	800	11.4	2.7	7.8								
8	3.6	.80	2.88	41.52	.007	800	9.6	3.0	8.6								
6 (one half)	1.8	.80	1.44	42.96	.005	630	9.0	3.0	4.3								
9 (one half)	1.8	.80	1.44	44.40	.007	630	7.6	3.2	4.6	.004	19.1	2.1	93.3	52"	400		3d St., Ave. C to Ave. D

* r and R taken from the rainfall-curve (Plate IV) for the New England States.

amount of run-off from all the drainage-areas above $= AIR$. $q(= aIr)$ as well as Q should be calculated for each sub-area, and if the Q for any stretch of sewer is at any place less than the q immediately tributary to the same the latter should determine the size.

Plate IV will be found convenient for determining a, and also R when the rates of rainfall of the place in question can be represented by any of the curves there given, these being for rains of the second class. To find a in acres from the diagram, use one dimension (in feet) of the area (or of an equivalent rectangle if it is not rectangular) as an ordinate and find the corresponding abscissa of the acre-curve in the diagram; divide this into the other dimension of the area and the quotient will be a in acres.

By the table the run-off from the undeveloped territory is placed at 55.2 cubic feet per second, which is carried by a 42-inch sewer on a .6% grade for 360 feet, where it receives still more sewage, the maximum amount to be received there, both over the surface and through the sewer, being 57.2 cubic feet, although q for the block No. 1 alone is 6.8 cubic feet. But Q is not equal to $55.2 + 6.8$, because the latter quantity was due to a rainfall of 6.25 minutes' duration, or rather to the maximum rate for that time, during which only such water would have arrived from the upper end of the drainage-area as was due to a lower rate of rainfall; but the time of 16.9 minutes is that for which the run-off is calculated from both the undeveloped territory and block No. 1. Blocks No. 2 and No. 3 both reach the sewer at the same point, and, taking the rate of rainfall for 17.7 minutes, we have a total run-off from all the territory above of 66 cubic feet per second and, the grade being .5% a 44-inch sewer is found to be necessary.

Blocks No. 4 and No. 7 discharge first into a branch sewer, which it is found should be 24 inches in diameter.

Where this joins the main the run-off from blocks No. 5 and No. 8 and from half of No. 6 and No. 9 also reaches it, and it must consequently be increased in size. The time T at this point is $16.1 + 0.8$ (in the Ave. B sewer) $+ 0.8$ (in the Second Street sewer) $+ 1.4$ (in the Ave. C sewer), or 19.1 minutes, and the rate of rainfall for this time is used for the run-off from the entire area.

It will be seen that the method here employed is but a practical application of the principles stated in Art. 18. More, and more accurate, data for determining t and I, as well as R, are needed before this or any method can be relied upon to give more than general approximations to the run-off. Fortunately with the method here given the approximation becomes more close as the area becomes more urban, and is most so in the most densely populated districts, where the danger from gorged sewers would be greatest.

ART. 37. GRADE, SIZE, AND DEPTH OF SEWERS.

For both determining and recording the grades of the proposed sewers use is usually made of the profiles of the streets, plotted from the level notes. Upon these a vertical longitudinal section of the proposed sewer through its centre line is placed, thus showing the size, grade, and depth of the sewer. While designing, however, it will be found convenient to pencil in the line of the invert only, since then changes in its vertical location can more readily be made.

A short experience in sewer-designing will demonstrate how mutually involved are Q, S, the diameter and the depth of the sewer. In many cases it will be necessary to alter and realter the grade and diameter before obtaining for each reach of sewer the best obtainable depth and velocity. Q is a fixed quantity for any given case, S may vary between fixed limits, the size also has its limits in some cases, but the depth of the sewer may vary from any distance below to any distance above ground. A depth of 25 or 30 feet is obtained in many sewerage systems, and even 50 feet or more has been reached in open cut, while sewers have been laid in tunnel at still greater depths. Where possible deep sewers should run through wide streets, that the danger to building-foundations may be kept as small as possible; and they should avoid the busiest thoroughfares unless these are also the widest streets and the soil is treacherous. The sewer may in some cases be carried on bridges or trestles, as in crossing a stream or ground lower than the hydraulic gradient. In many such locations, however, this position will be impossible, owing to traffic on the river, to danger from floods, to blocking of streets, or to prohibitive cost of construction. In such cases the pipe may be placed under the surface of the ground or in the bed of the stream, being thus below the hydraulic gradient. Such a downward loop is called an inverted siphon. Many instances of these are in use, and if care be taken in their design and construction they need give no trouble. It will not be possible, or at least advisable, to connect any buildings to an inverted siphon, since the sewage will continually stand in the connections up to the level of the hydraulic gradient of the siphon.

The depth of storm-sewers is usually fixed by grade requirements only; the covering over them, however, should be not less than two feet and would better be three or four feet. The minimum depth to which house or combined sewers should be laid will usually be decided by local circumstances or customs. It is generally desirable to lay them somewhat deeper than the gas- or water-pipes, that these may not interfere with them. The city of Brooklyn some years ago fixed 12 feet as the depth to which all (combined) sewers are to be laid, unless the maintenance of proper velocity requires a less or greater one. In Philadelphia 14 feet is the standard depth, in Washington, D. C., 10 feet. In residence districts in the smaller cities 7 to 10 feet is usually sufficient, although in a street running along a hillside a much greater depth may be called for by the depth of basements upon the lower side of the street. In streets which are already built up the sewer should be deep enough to drain all basements and cellars, with the exception, perhaps, of an occasional one of unusual depth. To insure this the cellar depths taken during the survey should be indicated in their proper positions upon the profiles of their respective streets. In many Southern cities where there are no cellars under the dwellings and there is little danger of frost the sewers may be given a depth of covering of only 3 or 4 feet. In the North 6 feet is probably the least depth which should be given to the flow-line save under exceptional circumstances. The maximum depth should be kept at 14 to 16 feet if possible, since below this the cost rapidly increases. When the depth is considerable the expense of making house-connections may become excessive. It may in such cases be found cheaper to lay a small sewer about 7 or 8 feet below the surface and following the surface grade, which may be with or against the grade of the deep sewer, to a manhole in the deep sewer into which this shallow one can discharge.

Before fixing the grade it is well to prepare tables similar to those given in Articles 35 and 36, and also to calculate as closely as possible the total amount of sewage reaching each outlet.

Very often the main sewer for a long distance from the outlet must be laid at a minimum grade if pumping is to be avoided or the lift kept as small as possible. In such a case the grade of this main will be the first to be located upon the profile, the outlet being placed as low as is permissible. This should never be below ordinary high water unless absolutely necessary, and under no consideration should it be below or even as low as ordinary low water; or rather this should be true for the hydraulic gradient, although the last few feet of the sewer may be given a steeper grade to bring the outlet below the water-surface or into the channel.

There may be other lines also where the surface elevations demand the flattest possible grades; that is, the grades which will give the minimum permissible velocity. This grade will depend upon the size of the sewer, and this again upon the quantity of sewage. To ascertain this size, reduce the maximum sewage flow to cubic feet per second, divide by the desired velocity of flow in feet per second, multiply the quotient by 1.5 for mains or 2 for laterals, and find the diameter of a circle having this product as its area, which will be the sewer diameter required. Or, divide the gallons of sewage per day times 1.5 or 2, as the case may be, by the required velocity in feet per second, and take the size corresponding to the next highest quantity in the following table:

TABLE NO. 16.

Size of sewer	8″	10″	12″	15″
Gallons of sewage per day $\frac{}{v}$	225,500	352,500	507,600	792,800
Size of sewer	18″	20″	24″	
Gallons of sewage per day $\frac{}{v}$	1,141,700	1,410,300	2,030,500	

Where possible the grades of house-sewers should be such as to give a velocity of from 3 to 4 feet per second, and those of storm-sewers from 4 to 5 feet per second. The demands of economical construction and the necessity for sufficient fall in house-connections should not, however, be sacrificed to reduce velocities to less than 10 or 12 feet, which, however, should be the maximum allowed.

If it is possible the grade of the various sewers should be so proportioned that the velocity of the sewage shall increase as the outlet is approached, or at least it should not decrease, since a decrease in velocity may cause a deposit of suspended matter. Frequently, however, it is impossible to attain this in the design, since the flattest surface slopes are usually nearest to the outlet and the sewer grades are largely controlled by these.

From the formulas $Q = aV$ and $V = c\sqrt{RS}$, considering the sewer as flowing full, and giving c a constant value of 85, which will in no case vary more than 10% from Kutter's c for pipes ranging from 8 to 18 inches diameter, we have the formula

$$V^5 = \frac{c^4 S^2 Q}{4\pi} = .0796 c^4 S^2 Q,$$

or

$$V = 21.1 \sqrt[5]{S^2 Q},$$

V being in feet and Q in cubic feet per second. This formula should not be used where any considerable accuracy is demanded, but will be found convenient for use in fixing the first approximate grades. If V is to be a constant S must vary inversely as $\sqrt[5]{Q}$.

S, however, equals $\frac{f}{l}$ if f equals the fall of the grade for a length l. If l, l', l'', etc., be taken as the lengths between successive manholes, f, f', f'', etc., as the corresponding falls, and Q, Q', Q'', etc., as the quantities of sewage flowing through these lengths, then

$$\frac{f}{l}\sqrt{Q} = \frac{f'}{l'}\sqrt{Q'} = \frac{f''}{l''}\sqrt{Q''}, \text{ etc.};$$

also $f + f' + f'' +$ etc. $= F$, the total fall from the head to the outlet of the system. Knowing F, and l and Q for each length between manholes, we can obtain the values of f, f', f'', etc. As just stated, it is seldom that an entire system can be designed to give a constant velocity to the sewage, but this is sometimes possible in separate drainage-areas, a constant velocity being obtained in each area.

Still more important than obtaining a constant or constantly increasing velocity is the keeping of the velocity within the limits given in Art. 22. If the ground-surface is too flat to permit of obtaining this velocity by gravity pumping must be resorted to (see Art. 42). If the surface is steeper than is permissible for the sewer the sewer grades can be broken and a drop made at each manhole (see Art. 41).

A slight drop in the grade should be made at each manhole on flat grades to compensate for the obstruction offered by curves, etc., at this point, and for slight errors in measurement. 0.02 or 0.03 feet is usually sufficient.

Of the above principles the most important is that the velocity of the sewage shall be within the proper limits; then that all basements and cellars to a reasonable depth shall be drained by house-sewers; also the depth of excavation should be kept as light as possible, and the principles outlined in Art. 34 should be regarded. The obtaining of the nearest possible approach to an ideal design will usually require many changes in, and rearrangement of, both lines and grades, since a change in those of one lateral may in some way affect the entire system.

The preliminary grades having been thus fixed according to the desirable depths and velocities of flow, the size of the sewer for each reach should be calculated or taken from the diagram and the velocities checked by accurate calculation. Additional changes, usually slight, will probably be required to obtain the best values for each interdependent velocity, size, and depth. The junctions and crossings of the sewer-lines must be carefully examined and adapted to each other. It is a good plan to make a list of all the manholes, showing for each the elevation at which each sewer enters and leaves it. Two sewer-lines should never intersect each other, each having a continuous grade; either one should discharge from both directions into the other or they should cross, the one above the other.

At junctions the surface of the sewage in the contributing sewer should never be designed to be lower than that in the other; that is, if they are both branch sewers the centre of the tributary should not be below the centre of the intercepting sewer; if the larger is a main the centre of the smaller should not be lower than a point two thirds the diameter of the larger above its invert. It would be still better to place the invert of the tributary above the sewage-surface in the interceptor, particularly when the former drains but a small district; but where the total fall possible is slight none of it need be utilized for this purpose.

Difficulty will sometimes be found in so arranging the comparative depths of storm- and house-sewers that the house-connections can pass under or over the former. In some cases this may be impossible, and it may be necessary to place a house-sewer on each side of the storm-sewer.

Reference to the data of locations and depths of gas- and water-pipes and other existing sub-surface systems should be constantly made and the sewers so designed as to interfere with them as little as possible.

On the profile of each sewer-line the elevation of all transverse sewers should be indicated and a cross-section of the sewer shown. On the finished profile it is well to indicate the thickness and material of the sewer-walls and of all manholes, lamp-holes, and other appurtenances. The materials may be indicated by colors, as red for brick, brown for sewer-pipe, etc. The grade, length, and size of the sewer between each two manholes should be given in figures, as well as the exact elevation of the invert at each change of grade.

Art. 38. Inverted Siphons.

Since the ordinary sewer is designed to flow only $\frac{1}{2}$ to $\frac{2}{3}$ full, while an inverted siphon, being under a head, will flow full bore, the velocity in the latter will be only $\frac{1}{2}$ to $\frac{2}{3}$ that in the sewer laid to the hydraulic gradient, if they are of the same size. On account of the difficulty of access and repairs it is especially necessary that the velocity of flow in the siphon should be at least as great as that in the ordinary sewer, that deposits may be prevented. This can be attained only by reducing the size of the siphon-pipe. Moreover, this velocity should be had from the beginning of the use of the system; and therefore this size should be designed to give sufficient velocity to the sewage from the first. This first sewage flow may be doubled or trebled as time passes, and the increase may then be provided for either by giving sufficient fall to the siphon originally to produce the greater velocity necessary or by additional siphon-pipes. Usually at least two siphon-pipes are laid at the first, that while one is being emptied and cleaned the other may be used. The friction-head in the inverted siphon will be greater than if the sewer were laid to the hydraulic gradient, and consequently the gradient must be steeper. The difference in elevation of the two ends of the siphon should be equal to the fall required by a sewer of the same size flowing full and of the length of the entire siphon (which is not the horizontal distance between its ends) to pass the given amount of sewage.

The velocity of flow in an inverted siphon is entirely independent of the fall therein, but depends upon the quantity of sewage, since all of this must, but no more can, pass through it. If the fall in the inverted siphon is not sufficient the sewage will back up the sewer until sufficient head is obtained to produce the required velocity. Hence to prevent this the fall in the siphon itself should be made great enough to create the velocity which will be required by the largest quantity to be passed at any time.

An inverted siphon may at times be necessary for passing under some obstruction in the street—as a large conduit of one kind or another, but this should be avoided where possible.

For details of inverted-siphon construction see Articles 49 and 77.

Art. 39. Sub-drains.

Very frequently storm-sewers are placed at such a short distance from the surface that they cannot be utilized for draining damp cellars, particularly since a cellar should be connected with no sewer whose *crown* is above its level, from danger of back-water when the storm-sewer flows full. Ordinarily the house-sewer is below the cellar-level; but this should not be utilized as a drain, both because the amount of sewage may thus be too largely increased; and still more on account of the danger from sewer-air, which would have free access through the drain should the trap-seal evaporate during a drought, which it is very apt to do, and from the cellar this air might permeate the entire house.

From a sanitary point of view the drainage of wet soils is almost, if not quite, as important as the sewerage and should not be neglected. The mere opening of sewer-trenches tends to drain the soil, even after they are refilled. But in many cases it is extremely desirable to provide other and more positive drainage.

It is almost impossible to make a perfectly tight sewer without great expense, and when laid in wet ground sewer-joints may admit in the aggregate large quantities of water. This could be prevented and the land adjacent drained, to its great improvement and the health of residents thereon, if this ground-water could be lowered along the trench by some means.

During construction in wet ground much trouble will be experienced, even when the pumping facilities are ample, by water rising and flowing over newly laid inverts, to their permanent injury (see Arts. 75 and 76).

These difficulties can each and all be met in most cases by the use of sub-drains—that is, drains laid a little below the sewers. These are ordinarily laid in a narrow trench in the bottom of, and at one side or in the centre of, the sewer-trench. Their use for construction drainage will be considered in Part II. When properly designed for this purpose their size will in most cases be sufficient for the continuous drainage of the land and also for cellar-drainage. The instances will be very few, however, in which any approach to an accurate estimate can be made of the amount of sub-drainage which will be required in a system. But provision should always be made for sub-drainage wherever the soil is wet, for permanent drainage if for no other purpose.

The water flowing into such drains must have some outlet, and the most natural course would be, when the sewage is disposed of by dilution, to place the outlets of sewers and sub-drains at the same point. It may happen, however, that the necessity for sub-drains is not foreseen when the sewer-outlet is being built; or the place where they will be necessary may be so far from this outlet that a great length of otherwise useless drain-pipe must be laid to reach it; also the amount of ground-water may be so much greater than was anticipated, in spite of all investigations, that the drain-pipe near the outlet will not carry it all. In any of these cases another outlet may be desirable or necessary. This can frequently be found by leading the sub-drain in a special trench to a near storm-sewer or natural watercourse. In some cases, however, special means must be resorted to, such as one of the methods of pumping (see Art. 42).

If the sub-drain is necessary for construction purposes only it may be led to a sump-well where a pump is stationed, and broken and sealed at several points after construction is completed. (This last will be necessary, as otherwise the drain would continue to lead the ground-water to this point, which might become permanently and dangerously water-soaked.)

Although the sub-drain is in most cases smaller than the sewer, it must be laid at practically the same grade. The objection to flat grades in house-sewers does not apply to these so urgently, however, since the water flowing through them, after construction is completed at least, is usually free from suspended matter likely to cause deposits. The size and position, then, are the only elements of the general design to be decided upon. The size it will not be advisable to make less than 6 inches at the outlet or for long stretches, but for stretches of a few hundred feet only and through ground but moderately wet 4- or even 2-inch pipe may be used. Pipe larger than 10 or 12 inches is seldom used in any but exceptional cases. If a larger would be required (and instances can be named where the sub-drainage from a small town would more than fill a 36-inch pipe) special methods may be employed; such as dividing the sub-drainage system into small sub-systems, each having its own outlet, which may, when constructed under a storm-sewer, discharge into the sewer immediately above it or which may be at a near watercourse.

ART. 40. HOUSE- AND INLET-CONNECTIONS.

The connections between the sewers and opposite houses and storm-water inlets are of an importance second only to the sewer-mains. Any defect in one of the connections, while limited in the range of its effect, is fully as detrimental within that range to the proper working of the system as a defect in the main itself. Since the house-connections are subject to extreme fluctuations of discharge and hence to stoppages, as also to the formation of grease deposits, it is desirable that they be equally as accessible as sewer-mains for both inspection and cleaning, and also that their grade and alignment be given equal care in both the design and the construction. They should, if possible, be given a uniform grade of not less than $2\frac{1}{2}\%$. Where the house sits back from the street an observation-hole (see Art. 47) should be placed at the fence-line, and one should be placed wherever there is a change in the line or grade. There should also be a hand-hole in the pipe just after it enters the cellar. The junction with the sewer should be made by means of branches, either Y or T. It should never be made in pipe sewers by breaking a hole into the shell and inserting a pipe. If the sewer be larger than 20- to 24-inch a T is advisable, both because this offers easier inspection of the house-connection from its lower end, which inspection can be made by a person entering the sewer, and because the branch can be placed entirely above the ordinary level of the sewage, which position it should occupy when possible so as to cause no interference with the sewage flow. When the sewer is too small to admit a man, which size will also not admit of raising the branch entirely above the ordinary sewage flow without giving it too steep a pitch, a Y branch is preferable, because this will retard the flow less than a T, and because the house-sewage will enter the sewer at a less angle with its flow. The vertical angle which the branch makes with the horizontal should not ordinarily exceed 45° in small sewers, because of the interference with the flow and of the splashing caused by a vertical drop of sewage into their relatively small stream, and because of the danger that the weight of the house-connection may break in the crown of the sewer.

It is well to so place the branch in brick sewers that a trickling discharge from it will flow over the brick for the least possible distance, that deposits from such discharge may be avoided. In the case of combined sewers this would call for placing the branch but a short distance above the invert, but it should be given such a grade as to bring it higher than the crown of the sewer when it reaches the cellar.

Some engineers always use T branches, more always use Y branches, for house-connections; but the practice here recommended seems to best utilize the advantages and avoid the disadvantages of each.

The connections with inlets should never enter the sewer at an angle with its axis greater than 45°, on account of the great disturbance to the flow which would be occasioned. Where possible, and particularly in small sewer-mains, a manhole should be placed where each connection enters the sewer and the connection continued by a curved invert in the bottom of said manhole (see Plate VIII, Fig. 5).

It is difficult to calculate the proper size for a storm-water connection, but, since there is little disadvantage in having it larger than is actually required, while the effect of too small a pipe may be disastrous, it is advisable to make the size fully ample to discharge all the run-off from the heaviest storms. A 12-inch pipe is probably the smallest which should ever be used; while a 24-inch may be required if the sewer lies near the surface (thus giving little fall to the connection) and if the tributary area is large. Where considerable undeveloped territory drains into the head of a sewer-main, or a small stream is there received, it may be necessary to continue the sewer to the inlet, not only not diminished in size but even enlarged into a bell mouth. It would be advisable to use an increaser at the upper end of every inlet-connection, since, owing to the churning of the water in the inlet, a "standard orifice" will not pass more than two thirds the water which can be carried by a pipe of the same size.

ART. 41. MANHOLES, INLETS, FLUSH-TANKS, ETC.

The necessity for frequent connections between the air of the sewer and the outer air has been shown (Articles 28 and 29). As one means for this, and one which can always be adopted, manholes should be adapted to serve this end by having perforated covers. For this purpose, also, the more numerous they are the better. The other and greater necessity for their use, that of providing access to the sewers, should, however, have greater weight in fixing the distances which should separate them. It has been found in practice that a 6- or 8-inch sewer can be easily inspected and cleaned if this distance be not greater than 300 feet; a 12- or 18-inch sewer, when not more than 400 feet separates successive manholes. A sewer which can be entered may, for this purpose, have its manholes even 600 or 1000 feet apart; but the cost and difficulty of cleaning are thereby increased, owing to the distance the material removed in cleaning must be carried through the sewer. Ventilation also is not so well served by so great intervals. It is better to fix 500 or 600 feet as the maximum distance between manholes on lines of the largest sewers.

Economy would suggest placing a manhole at each sewer intersection, where it would serve both lines. This is also desirable as permitting a curved junction between the sewer-channels. Where a curved bend is made in the entire sewer a manhole should be placed at each end of the curve unless the sewer is sufficiently large to be entered.

A manhole should be placed, in general, at each change

of line or grade, in order that every part of the sewer may be easily inspected.

Economy will set a limit to the number of manholes which may be introduced; the number of the breaks in the street-paving caused by their covers it is also desirable to keep at a minimum. Principally for the first reason a manhole is sometimes omitted in small sewers when it would come less than 200 feet distant each way from another manhole, and a lamp-hole substituted. While the sewer cannot be inspected from this, a light can here be lowered into it to light up the sewer for inspection from the next manhole either way. Also a hose can be inserted at a lamp-hole for cleaning the sewer.

The use of flush-tanks has already been discussed (Articles 25–27). The grades of the laterals and the conditions of their use should be carefully examined to determine where frequent flushing will probably be needed. In some cases, such as where a flat grade on a long line of small sewer is unavoidable, it may be desirable to place automatic flush-tanks at intervals of 800 to 1000 feet along its length, the tanks being placed at one side of the sewer and discharging into it through a short connecting-pipe. If automatic appliances are not employed no special tanks need be built in such a case, but manholes at intervals along the line can be used for flushing.

All the local conditions should be examined that advantage may be taken of any opportunities for flushing offered by springs, streams, or any available sources of water, and in general decision made as to the places and methods of flushing. As a general rule every dead end of a house- or combined sewer should be flushed frequently and some arrangement for this placed at each such point.

Inlets should be provided at frequent intervals throughout the area drained to receive the surface-water. In districts where the street traffic is considerable and where any great depth of water in the gutters would inconvenience a large proportion of the population the inlets should be not more than 200 or 300 feet apart, while in residence districts they may be so situated as to require the run-off to flow for 600 or 700 feet over the surface. They should generally be so placed that all the run-off can reach them by flowing along the gutters only, and need not flow across the streets. The plan Plate V shows how this can be accomplished in most cases. Where this is impossible a culvert should be placed under the street-pavement in line with the gutter.

Where street grades are continuous from one intersecting street to another inlets should be placed on street-corners. They are frequently placed at the gutter intersection; but a better plan in many cases, particularly on steep grades, is to place two openings, one just above each cross-walk, as this avoids the vehicle-trap caused by the ordinary corner inlet. Also an inlet should be placed at every point where two falling grades meet, and if this be between street intersections an inlet must be placed there on each side of the street.

In the majority of cities a large proportion of the inlets are provided with catch-basins—more than the best practice would warrant, in the author's opinion. The object of using a catch-basin is to retain there the silt and other heavy matter and not permit it to be carried into and deposited in the sewer. Catch-basins should be cleaned after every storm.

The objection to catch-basins is that several days sometimes must elapse—and several weeks usually do—between the beginning of a storm and the cleaning of the catch-basin; and during this time the organic matter which has been washed or thrown into the inlet, including horse-droppings, fruit and vegetable refuse, etc., is putrefying and frequently emitting objectionable odors. " Such foulness is less offensive in the drains [storm-sewers] than in the catch-basins, which are situated at the sidewalks and where it is much more likely to be observed. Also it is found impracticable to intercept all matter in the catch-basins which would deposit in the drains after they reach the flat grades in the lower part of your city. The cleaning of the drains would, therefore, be necessary in any event, and the additional amount of silt that would be intercepted by the catch-basins will not cost much more to remove. In the city of Paris, even though a combined system of sewers is used, it is not found objectionable to allow all the street-dirt to enter the sewers and therefore the catch-basins at the inlets are omitted." (Report of Rudolph Hering and Samuel M. Gray on Sewerage of Baltimore.)

As a matter of fact catch-basins are not infrequently left uncleaned after light storms, or even heavy ones, for weeks together, and the odors from them are usually attributed to the sewers, which in most cases are far less foul. Moreover, catch-basins are usually cleaned with shovels only and sufficient filth left upon the sides and bottom to become noticeable by its odors. When cleaned so infrequently the catch-basin often stands full of material and is until cleaned practically non-existent so far as any useful effect is concerned.

For these reasons the universal use of catch-basins is, in the author's opinion, not to be advised, but rather the inlet should be so designed that all material shall at once reach the sewer. The inlet-connection he would also make without a trap, that it may assist in the ventilation of the sewer; and if the sewer and its appurtenances are properly designed, constructed, and maintained there will be very few instances where any odor can be detected at the inlet.

There may well be cases where catch-basins are desirable, as where the wash from a steep hillside is caught, or for other reason a large amount of coarse soil or " clean dirt " finds its way to the inlet; and there the catch-basin will need to be large, that but a small proportion of this may reach the sewer, and should be cleaned after every heavy shower. A small catch-basin is in most locations worse than useless.

Catch-basins are also desirable where the sewer grades are very flat and the velocity is less than 3 feet per second; also on combined sewers where the streets are unpaved.

ART. 42. PUMPING OF SEWAGE.

There will frequently occur instances where, even if the sewers be laid at the flattest permissible grades, either the outlet will come too low, or the upper ends or some intermediate point will be too high for proper service. This is especially likely to occur where the outlet is at a considerable distance from the city; also where treatment of the sewage is necessary. Under such circumstances there is but one solution of the difficulty—the sewage must be raised at some one or more points from a low to a higher level. (Where a street has not yet been graded or built upon it may often be practicable to lay the sewer above the ground-surface in crossing a valley or basin, and so grade the street finally as to give it a proper covering, thus avoiding the necessity of pumping.)

Where the sewage is discharged into tidal waters and the outlet is below high tide the lower stretch of the sewer will be filled twice a day, and the velocity therein cannot then exceed the quotient obtained by dividing the volume of sewage by the area of the sewer. It would therefore be well to make this sufficiently large for present needs only and duplicate it when greater capacity becomes necessary. In some instances tidal basins are constructed, which are closed —automatically in most cases—against the rising tide, and receive and hold the sewage flow during high tide, their contents being discharged on the falling of the tide. In some cases the sewers themselves are made sufficiently large near the outlet to serve as reservoirs in the same way. But these reservoirs are seldom satisfactory, owing largely to the difficulty of cleansing them from the deposits made while they are filled with stagnant sewage. It would be better, though of course more expensive, to pump the sewage during high tide; or better still to raise the streets and sewers generally, where this is possible, and discharge above high tide. (The city of Chicago some years ago raised the streets over its entire area to permit of better drainage.)

In certain places the conditions are such that the water rises above the sewer-outlet, which is ordinarily free, for periods of days or even weeks; as on a lee shore during a storm or on rivers subject to extended floods. In such a case pumping is necessary; but the first cost of the plant should be kept at a minimum, since the interest on this will far exceed any saving that could be made in running-expenses for a few days. If possible it is well to locate the plant where power can be obtained from an outside source—as steam from the boilers of a water-works pumping-plant, electricity from a power or traction company, etc.—by which means both first cost and running-expenses may be reduced.

Where house-sewage only is to be raised the apparatus should be of a capacity sufficient for the maximum flow. Storm-sewage, or at least the entire run-off from heavy storms, is not often pumped, owing to the enormous capacity required in the machinery. It will in most cases be found more economical to build special outlets for the storm-sewage to the nearest watercourse, where this is practicable. In the case of a combined sewer the house-sewage should all be pumped, as should even the run-off from light storms, which carries street-washings. But it will usually be permissible to allow the run-off from heavy rains with the admixture of house-sewage to escape by overflows and special storm-sewers to nearer outlets. If this would give rise to danger or a nuisance, owing to even the small proportion of house-sewage contained, it is probable that the separate system should be employed, all house-sewage being pumped and each storm-sewer seeking the nearest outlet.

In a very flat country it may be desirable to raise the sewage at a great number of points to prevent deep and expensive excavation. A sewer under a level surface, beginning at a depth of 8 feet and falling 1 foot in 300, would in 2100 feet have a depth of 15 feet. Beyond this the cost of construction would rapidly increase unless the sewage could be lifted and started again at a depth of 8 feet.

Whether the lifting of the sewage shall be done at one station or at several is usually a question of cost only. It can be exactly settled only by a comparison of the sum of the interest on first cost and the operating-expenses of one method as compared with another. (It is assumed that the depth of every sewer is made sufficient to meet all requirements.) The fewer the lifting-stations and the further apart they are the greater will be each lift; also the greater will be the average depth of sewer. Hence, while the greater the distance between lifts the less will be the total cost of lifting machinery or apparatus, and also of maintenance of the same; on the other hand the greater will be the cost of the construction of the sewer and also of its maintenance. The proper decision as to the number and location of the lifting-stations is frequently a problem requiring much careful study. While in one locality, where excavation is expensive, 5 feet may be the maximum lift which will be economical, in another this limit may reach 30 feet or more. If all the lifting can be done at one or two points it is usually most economical to so arrange it, even at great expense for excavation.

The methods and apparatus to be employed may be: pumping by steam, gas, gasoline, or hot-air engines or electric motors, lifting by a Shone Ejector, an Adams Sewage-lift, or other appliance which seems adapted to the circumstances.

If steam, gas, gasoline, or hot air be employed a complete plant must be placed at each lifting-station. Where electricity is the motive power a motor and pump only are required at each station. This renders possible a saving by using electricity, under certain conditions, such as many lift-stations with a small horse-power required at each, or even when the horse-power is considerable. For five stations at New Orleans, of very great pumping capacity, B. M. Harrod estimates the annual cost as follows:

	Steam.	Electricity.
Interest and depreciation	$42,143	$58,684
Operation	69,960	54,825
Total	$112,103	$113,509

If the difference in cost of real estate for the two systems be allowed for, the annual cost would probably balance very closely.*

The pumps usually employed are the piston- or plunger-pump and the centrifugal pump. Other devices have been employed, such as screw and oscillating pumps, but few with any success. The centrifugal pump requires a quite constant volume of sewage for its proper working; hence, usually, a storage-basin, which is objectionable. For low lifts, however, it is frequently more economical than a piston-pump; also the wear due to grit in the sewage is neither so great nor so injurious to the pump, and hence the necessity for screening the sewage is not so great as with the piston-pump. With the latter particularly care should be taken to remove all large solids and gritty matter. For this purpose gratings, wire screens, and settling-tanks are employed, the last being of such cross-section that the velocity through them is less than one foot per second. These should be near or in the pump-

---

* Since the above was written electric pumping has been adopted for this work.

ing-station in order that they may be under the inspection of the engineer and that the deposits may be raised to the surface by power. If a steam-plant is used the screenings can be burned on specially prepared grates.

The Shone Ejector is a device for raising sewage which is actuated by compressed air. It is usually employed where a number of lifting-stations are needed, and the compressed air for all is supplied through iron pipes from one air-compressing station. While the prime motive power, steam, is employed indirectly, the efficiency of compressor, air-pipe, and ejector combined is greater than if a number of separate steam-pumps are used, with either separate boilers or a central steam-plant, especially when the stations are numerous and widely scattered. For only two or three stations the economy of their use is doubtful.

At Margate, England, sewage-lifts are used, with city water under considerable pressure as a motive power.

At Aberdeen, S. Dak., two Worthington motors connected with a sewage-pump are driven directly by pressure of the water from an artesian well. The capacity is 3,500,000 gallons per day, lifted 23 feet.

The Adams Sewage-lift can be employed where the surface grade at some part of the system will admit of introducing a drop, either vertical or on a steep grade through a pipe under pressure, in the line of some sewer or sewers. The sewage in making this drop transfers its energy by the medium of compressed air through pipes to a lift-station. The more frequent application of the Adams lift, however, is in flat districts where city water is usually employed for compressing the air, the supply being controlled by a ball cock in a catch-basin at the lift-station.

From none of these lifting appliances is there any odor, under good management. They can therefore be placed at any convenient point. The small pumping-plants, the Shone

Ejector and Adams Lift, are usually placed in vaults beneath the surface, the larger plants above ground. The sewage-pumping stations of London and Berlin are within the city limits, no odor whatever being perceptible near them.

ART. 43. INTERCEPTING-SEWERS AND OVERFLOWS.

It often happens that a town lies in a valley and upon the slope on one or both of its sides, and that while the valley district is too low to sewer to the outlet by gravity the upper districts are sufficiently elevated to do so. In such a case it would be useless to carry all the sewage to a main lying in the valley and raise it all to a gravity outlet-line. Instead a gravity-main should be run up each side of the valley at the minimum grade to receive all the sewage from higher up the hill, leaving only the sewage from below this to be pumped. Such a main is called an intercepting-sewer.

In some instances a combined sewer is provided with an outlet to the nearest watercourse, which is for storm-sewage only, it being intended that the house-sewage shall be received and conducted away by another sewer, which also is called an intercepting-sewer.

This term is also applied to a long sewer which passes down a valley and receives the sewage from several systems or parts of systems to conduct it all to a common outlet.

It is frequently advisable, when the gravity-outlet must be below high tide, to locate an intercepting-sewer which can discharge above all tidal influence, that the effect of the sealing of the lower outlet may be felt by only a part of the system, the upper sections discharging through the free outlet of the intercepting-sewer.

It sometimes happens that a system must be extended further in a given direction than was anticipated, or that the amount of sewage contributed by a district becomes greater than the sewers can carry. This can be remedied by running an intercepting-sewer across such gorged sewers at mid-length, intercepting the sewage from above and leaving the lower lengths to carry only their local sewage.

Where storm-water can find near outlets from many districts to a stream or other body of water, at which outlets, however, the house-sewage should not be discharged, an intercepting-sewer may be run along and near the water to intercept the house-sewage and convey it to a satisfactory outlet or to a disposal grounds or works. By a construction of the sewers called an interceptor (see Art. 48) the house-sewage and the run-off from light rains, which is the filthiest of storm-sewage, may be diverted to the intercepting-sewer, while the run-off from heavy storms will reach the nearer outlet. Mechanical contrivances for diverting the sewage are also used (see Art. 48).

Another method of obtaining similar results is that of putting storm-overflows in the combined sewers, a special storm-sewer taking the overflow sewage to a convenient outlet. The overflow is, in general, an opening in the sewer with its bottom elevated some distance above the sewer-invert. Until the sewage reaches the height of this overflow it remains in the combined sewer and flows to its outlet; when the quantity becomes such that the height of sewage flow is greater than this the surplus discharges through the overflow into the storm-water outlet. It is usually so arranged that this shall occur only when the dilution of house-sewage by storm-water has reached the point where the discharge of the mixture into a stream is free from all danger.

With either of these constructions the overflow or the interceptor should, if possible, be at such an elevation that it cannot be reached by floods or tides backing up the storm-water sewer.

Art. 44. Use of Old Sewers.

In many cities, before any general sewerage system is constructed or even thought of, short conduits, both private and public, have been built, discharging at the point nearest to hand—usually a stream or lake. These are often built in the crudest manner, graded by eye, and generally larger or smaller than necessary. In other cases the sewers are well built and graded and of a size adapted to remove the storm-water, but the outlet is located where house-sewage should not be discharged, or the sewer is not sufficiently deep to permit of receiving all house-sewage, or it is a pipe sewer and is not provided with sufficient branches for house-connections. Such sewers can frequently be incorporated into the proposed system, and a saving made of the cost and the tearing up of the streets avoided. But a thorough examination of them should first be made to ascertain which ones can be so used and how.

If they are sufficiently large they should be entered and their condition learned as to size, grade, character of workmanship, etc. If the brick-work is very rough it may be desirable to clean it and plaster it with cement mortar. It may be cleaned by washing first with dilute muriatic acid, then with a solution of potash, and then with water.

No connection-pipes should be allowed to protrude within the sewer. If the junctions are not well designed they should be torn out and rebuilt. If necessary a sufficient number of manholes should be built to bring the intervals between them within the proper limits. If it is desirable to use an old circular sewer as a combined sewer the invert can be narrowed as shown in Plate VII, Fig. 7.

If the sewers are too small to be entered they should be examined thoroughly from the manholes by means of mirrors (Art. 68); pills (Art. 85) should be passed through them to ascertain whether the bore is of uniform size and clear of deposits. Their size, grade, elevation, etc., should be learned by actual measurement. If they are not laid in straight lines, particularly those less than 12 or 15 inches in diameter, it is doubtful if they should be used, unless manholes and lampholes can be so judiciously located as to give straight stretches of sewer between them.

If a pipe sewer is too high for efficient service or at too flat a grade a trench may be sunk along its line and the pipe taken up, cleaned, and the good ones relaid at a lower level or better grade in the same trench. In the majority of cases this probably will be the best disposition which can be made of old pipe sewers.

Owing to the difference in character and volume of house- and storm-sewage a sewer not adapted for use as a house or combined sewer may often be used as a storm-sewer. It frequently happens that old combined sewers, or even the larger house-sewers, are admirably adapted to this use, and a separate system can then be built for the house-sewage.

If an old combined sewer, or storm-sewer modified into a combined sewer as explained above, can be used, except that the house-sewage should be discharged at a new and more distant outlet, this sewage can be discharged through an interceptor, or diverted by a mechanical regulator into an intercepting house-sewer, and the old outlet used to discharge the storm-water only.

But the efficiency of the system is of greater moment than small economies, or even large ones, and should not be sacrificed to them.

CHAPTER VIII.

DETAIL PLANS.

ART. 45. THE SEWER-BARREL.

SEWERS have been made of almost every conceivable shape and the walls built of all kinds of materials. A few shapes and materials are of almost universal applicability, others are adapted to peculiar circumstances only, and some are freaks of invention adapted to no circumstances.

The shape of cross-section is to a certain extent controlled by the material of which the sewer is constructed. The smallest sewers cannot be advantageously built of brick, but are usually composed of earthenware or metal pipes or of concrete. Earthenware sewers are made from 2 to 42 inches interior diameter. They are seldom made other than circular, owing to the liability of other shapes to become distorted in burning. Metal pipes are employed where the sewer will be under pressure, as in a siphon, or where there is a great deal of ground-water; also sometimes to better resist disturbing forces, as in made or treacherous ground or outlets under water or in shifting sands. The only metal commonly employed is iron. Metal pipes have always been made circular, although there are none but economic reasons why other forms could not be made.

Concrete and cement sewers are made of all sizes and shapes—circular, egg-shaped, rectangular, etc.—the smaller sizes being usually of cement, the larger of concrete.

Wooden-stave sewer-pipe has been used in the West, and in the East to some extent. On the Los Angeles outfall sewer are 34,100 feet of 36- and 38-inch pipe of this description. The outlet sewers in New York and Brooklyn are many of them creosoted wooden-stave pipe of 3 or more feet diameter.

For all sewers the circle is the most economical shape, and generally the most desirable, if they are never to run less than ⅔ full, except that the use of platform foundations may modify the first statement. But if they are to be used as combined sewers the egg shape is to be preferred, or a form similar to Plate VII, Figs. 2 and 6.

In Brooklyn, N. Y., and a few other cities cement sewer-pipe is used, and in general all sizes of this above 12 inches—in Brooklyn all sizes—are egg-shaped. Sections of this pipe are shown in Plate VI, Figs. 1 and 2. The flat base is given the pipe to prevent its rolling in the trench after being placed in position and to strengthen the bottom against crushing.

In the case of large sewers, particularly those whose diameter exceeds 4 or 5 feet, it frequently becomes necessary to make the width greater than the height, because the depth of the invert is limited by sewer-grade requirements and the height of the arch by the street grade. A great number of shapes have been designed to meet these conditions. Some of the best are shown in Plate VI, Fig. 5, and Plate VII, Figs. 9 and 10. Plate VII, Fig. 4, shows a design for very low head-room, but the thrust of the arch is considerable and the side walls should be heavier than shown unless they are firmly backed by rock or solid earth. Plate VIII, Fig. 1, is a better design to employ where the head-room can be slightly increased.

The use of steel beams for supporting the roof, with vertical side walls, as shown in Plate VII, Figs. 9 and 10, is becoming quite common, and is probably the best construction for soft ground with limited head-room. Fig. 10 is adapted to storm-water only, or to a flow of house-sewage never less than 15 inches deep. The egg-shaped sewer in Fig. 9 is intended for the house-sewage, the larger channels for storm-water.

Plate VIII, Figs. 2 and 3, show substitutes for egg-shaped sewers where the head-room is contracted. In Fig. 3 the semicircular invert should be sufficiently deep to admit of carrying the maximum house-sewage flow, that the sloping benches may not be fouled by it. Fig. 2 is especially adapted to an exceedingly variable house-sewage flow, as from a factory district whose Sunday and holiday flow is inconsiderable.

Plate VI, Figs. 5 and 9, Plate VII, Figs. 4, 5, and 10, and Plate VIII, Fig. 1, are best adapted to storm-sewage only, although they may be used as combined-sewer mains if the depth of the house-sewage flow is never less than 4 to 6 inches at the shallowest part, and the velocity is then sufficient. Plate VI, Figs. 1, 6, 7, and 8, are intended for house-sewage only. In Fig. 7 the flat invert is permissible owing to the constant depth of the sewage flow, which consists of intercepted house-sewage from a number of residence suburbs.

Plate VI, Figs. 2 and 3, Plate VII, Figs. 1, 2, 3, 6, 7, and 9, Plate VIII, Figs. 2 and 3, are intended to act as combined sewers. In Plate VII, Figs. 5 and 6, the side bench is horizontal, that it may serve as a sidewalk for sewer inspectors and cleaners.

The circular or egg-shaped form demands for strength a solid support under its invert. Where the soil is clay or firm loam, or a mixture of these with sand or gravel, or rock easily shaped, such a sewer may be built with walls of uniform thickness, the invert bearing upon ground shaped to receive it. If the ground is not firm, however, or cannot be readily shaped, the sub-invert spaces must be filled with concrete,

160 SEWERAGE.

Plate VI.

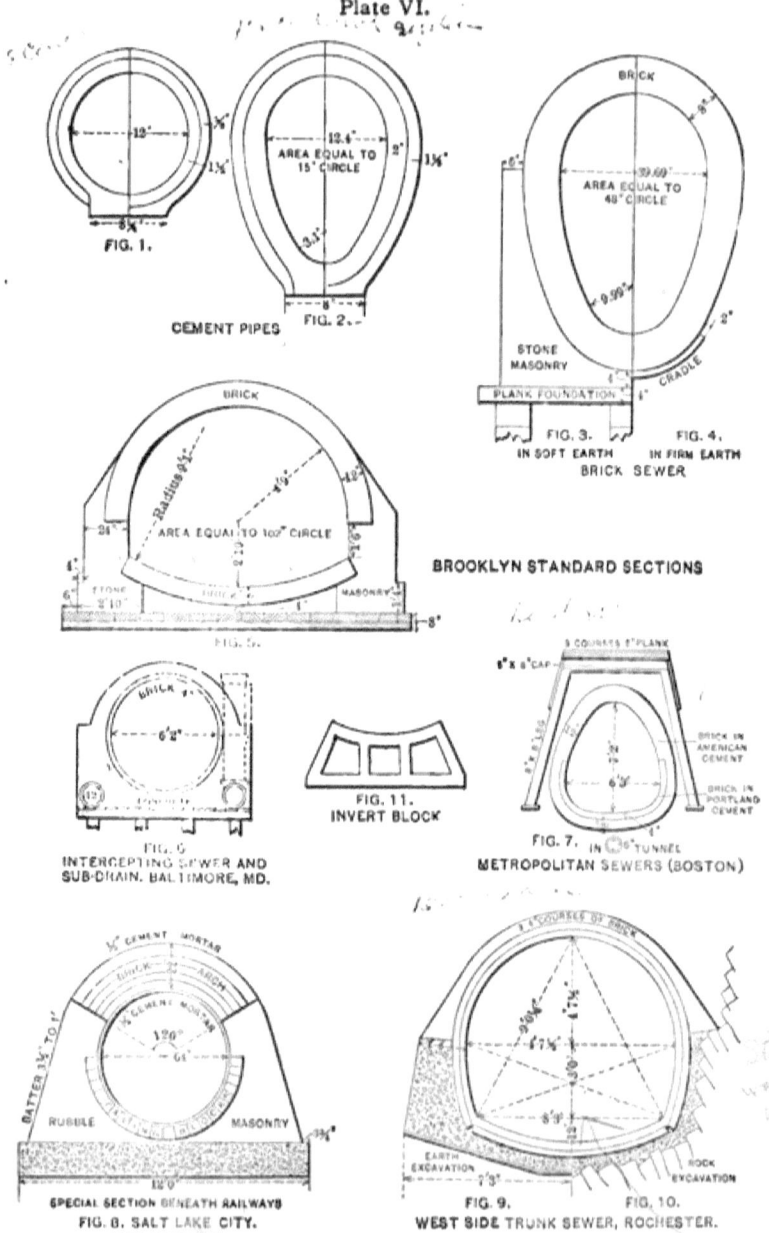

brick, or stone masonry, as in Plate VI, Figs. 3, 5, 6, 8, and 9. If the arch is of such dimensions that the horizontal thrust becomes more than the soil can receive without yielding, then the side walls must be designed to receive this thrust, as in Plate VI, Figs. 5, 6, 8, and 9. The general principles of arches apply, of course, to arched sewers, one of the most important being the necessity for stiffness of the haunches.

The circle, as has been stated, is the most economic shape for a sewer when the invert requires no backing. When this is necessary, however, the circle becomes an expensive shape, and the most economic is one with vertical side walls and bottom flat or conforming generally to the shape of the trench bottom. This is seen by an inspection of Plate VI, Figs. 6 and 8, Plate VII, Figs. 4 and 10. It is for this reason that most of the flat-bottomed sewers are built. Permanency of construction demands a covering for timber platforms, which are liable to abrasion and also to rotting away. This covering, forming the sewer bottom, is usually given a curved form, as in Plate VI, Fig. 5, or a sloping one, as in Plate VIII, Fig. 1, for two reasons: to concentrate small streams and decrease deposits, and to give strength to the bottom to resist the upward pressure which will exist when the soil is soft mud, quicksand, or similar material.

The materials of which sewers are commonly composed are brick, stone, and concrete masonry, cement and vitrified salt-glazed pipe, and, under special conditions, cast- or wrought-iron or steel pipe.

Stone and brick masonry is usually built up in cement mortar, and cement is always used for concrete. The stone masonry is usually rough, but compact and well-built, rubble. In arches brick is usually employed, as being cheaper and also stronger unless the stone are carefully dressed. The interior surface of the sewer, when this is built of stone, is usually

Plate VII.

FIG. 1. FIG. 2. FIG. 3.
WASHINGTON D.C. STANDARD SEWERS.

FIG. 4. TIBER CREEK, (WASHINGTON) SEWER IN 1893.

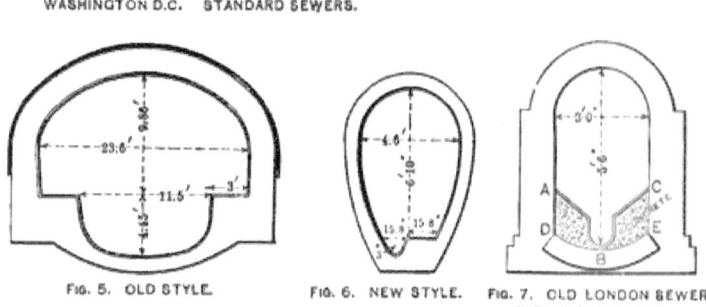

FIG. 5. OLD STYLE. FIG. 6. NEW STYLE. FIG. 7. OLD LONDON SEWER. WITH IMPROVED INVERT.

PARIS SEWERS

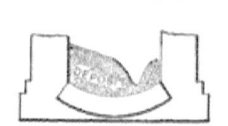

FIG. 9. DOUBLE STORM CHANNEL AND HOUSE SEWER; BRUSSELS. UNDER RAILWAY TRACKS.

FIG. 8. OLD LONDON "SEWER OF DEPOSIT."

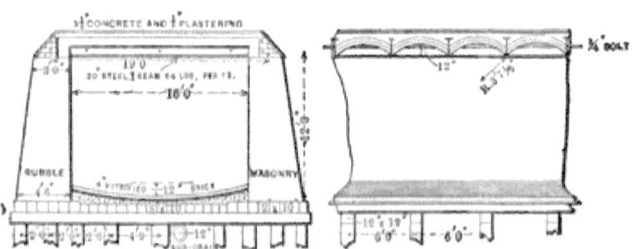

FIG. 10. CANAL STREET SEWER; ST. PAUL, MINN.

lined with a 4-inch ring of brick, because a brick surface can be more easily made smooth than can stone masonry (see Plate VI, Fig. 9). If much wear is anticipated smooth-dressed granite or trap blocks are frequently used as invert-lining (see Plate VI, Fig. 8).

Where the foundation is yielding a concrete base is frequently used under the sewer, as in Plate VI, Fig. 8, Plate VII, Fig. 9. But if it is soft a platform or even piles should be used under the concrete.

Sewers built entirely of concrete have been used in Europe very extensively and are coming into use in this country. In many localities concrete is cheaper than rubble or brick masonry. If well made a concrete sewer is both stronger and tighter than a stone or brick one, and can be made more durable than many kinds of stone or brick. The wearing-surface should be given a smooth coat of rich Portland-cement mortar $\frac{1}{4}$ inch to 2 inches thick, or a lining of hard brick, which is probably better owing to the liability of cement coatings to separate from the body of the concrete (see Plate VI, Fig. 6; Plate VII, Figs. 1 and 2).

If arches of small radius are built of brick-work laid with radial joints much cement is used, the arch is often weak, and the inner surface a polygon in section rather than a curve, unless brick especially shaped are used. If laid well such arches are also expensive in labor. To meet these objections, which apply particularly to inverts in egg-shaped brick sewers, invert-blocks of vitrified clay have been used. There are objections to these, the principal of which is that a joint entirely through the sewer is made, and where the hydrostatic head is greatest, which is almost sure to permit the leakage of water into or out of the sewer. They are also rather expensive, and are but little used now. A section of such a block is shown in Plate VI, Fig. 11.

A better plan for constructing short-radius inverts is by

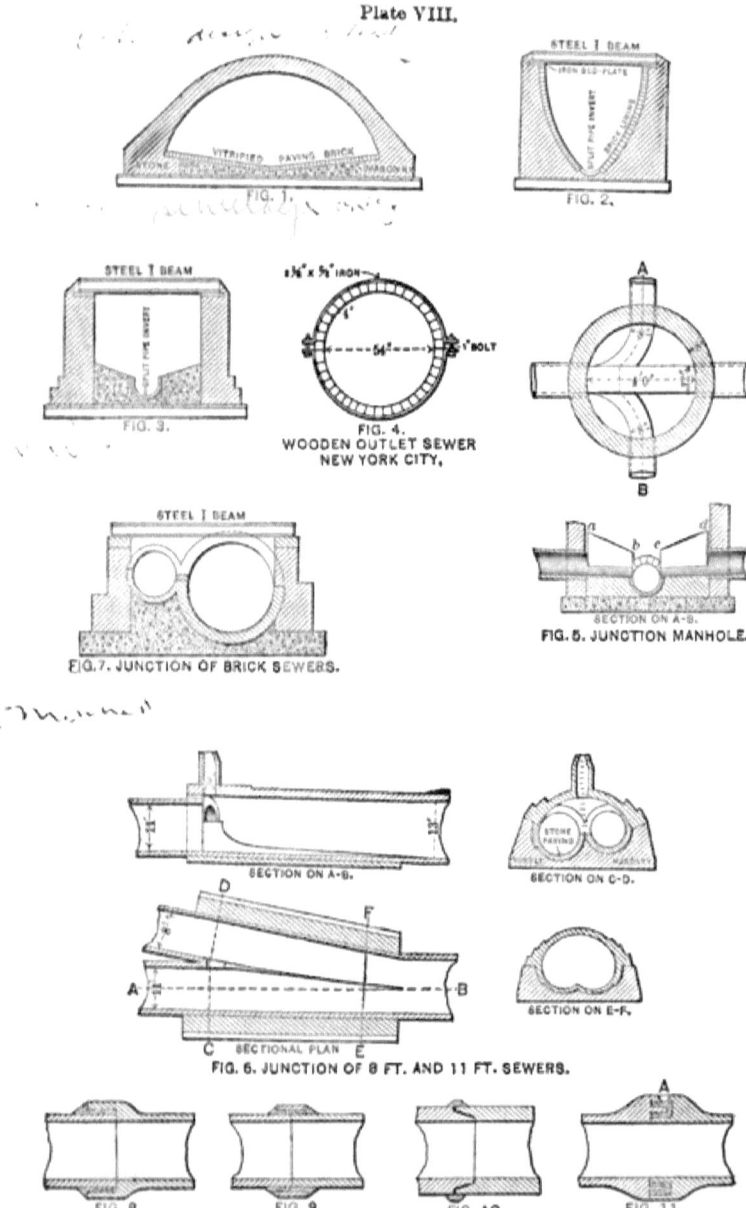

Plate VIII.

FIG. 1.
FIG. 2.
FIG. 3.
FIG. 4. WOODEN OUTLET SEWER NEW YORK CITY.
FIG. 5. JUNCTION MANHOLE.
FIG. 7. JUNCTION OF BRICK SEWERS.
FIG. 6. JUNCTION OF 8 FT. AND 11 FT. SEWERS.
FIG. 8. HUB & SPIGOT JOINT.
FIG. 9. RING JOINT.
FIG. 10. BEVEL JOINT.
FIG. 11. "ARCHER" JOINT.

the use of concrete or brick, lined on the inside with vitrified sewer-pipe split into thirds, which is approximately the arc of the small invert-circle in the egg-shaped sewer. Such a construction is shown in Plate VIII, Fig. 2. This construction is also well adapted to such sewers as are shown in Plate VII, Figs. 2, 6, and 7, Plate VIII, Fig. 3.

Whole vitrified pipe are used for lining to circular sewers up to 42 inches diameter, when the pipe is not used alone on account of the additional strength or tightness of joints required.

There is no fixed rule for the thickness of sewers, which depends upon the shape and diameter of bore, the material, the pressure received from the surrounding soil, and other circumstances. Brick sewers less than 30 inches diameter are frequently made but one ring—4 inches—thick; from this up to about 60 inches, 2 rings or 8 inches thick; from this up to 120 inches, 3 rings or 12 inches thick. This applies to the arch more particularly, unless the surrounding ground is very firm, when the invert may be made of equal thickness, or even 8 inches thick only when the arch is 12 inches or more thick. Some engineers never use less than two rings of brick in a sewer-arch; some use one ring up to diameters of 3 feet or more. The latter may give sufficient strength against crushing, but is hardly stiff enough to resist distortion except under unusually favorable circumstances.

The thickness of the side walls, when these are vertical, must be such as to enable them to withstand the pressure of the soil without or of the water within the sewer when it is full; also to receive the thrust of the top arch when the soil is not capable of doing so.

When two sewers intersect one or both should be curved in the direction of flow of the other. If one or both are small the curve may be made in a manhole (Plate VIII, Fig. 5). If one is many times larger than the other the curve may be omitted, the branch making an angle of 45° with the main sewer at the junction. Where they are each larger than 30 to 36 inches diameter the intersection should be made by bringing the two barrels gradually into one. This will require considerable skill in both design and construction when the tops and inverts are both arched. When the top is a girder construction the plan is much simplified, and still more so if the bottom also is flat. The crown of the sewer a short distance below the junction should be as low as that of the lower of the two sewers a few feet above it. A plan of a junction of two circular sewers is shown in Plate VIII, Fig. 6. If the head-room is limited the plan shown in Fig. 7 may be used. In Wilmington, Del., the junction of two 6-foot and a 10-foot sewer forms a chamber which is roofed with countergroined arches.

ART. 46. PIPE SEWERS.

Pipe is ordinarily used for sewers up to 18 or 24 inches diameter. Above this up to 42 inches vitrified clay pipe is sometimes used, but many engineers are doubtful of the strength of the larger sizes against crushing. The smaller sizes up to 18 or 24 inches, when made of good clay well burned, are sufficiently strong for ordinary locations, although the "double-strength" pipe (having a thickness of shell $\frac{1}{12}$ the diameter) is recommended rather than those of the standard thickness, which is less than $\frac{1}{12}$ the diameter by a difference which increases with the diameter. It has so far been found impracticable to make good, sound, symmetrical clay pipe with shells much thicker than $\frac{1}{12}$ the diameter. It is probable that if this thickness be maintained the largest sizes of pipe are amply strong for ordinary circumstances.

In many instances where vitrified clay pipe has been crushed in the ground it has been found that this was probably due to the fact that the pipe had a bearing on the bottom at only one or two points instead of along its entire length, or that stones or frozen earth were thrown upon it in back-filling. If earth is well tamped under and around a vitrified clay pipe it will not usually collapse, even when broken, although it may leak. Such pipe ordinarily breaks along four lines—at top, bottom, and each side—into pieces of almost equal size. For this reason fire-cracks and slight imperfections which do not cause the rejection of a pipe should be placed at a point about 45° above the horizontal in laying, and not at the top.

Several tests have been made of the strength of vitrified clay pipe. In one series, in which the pipe were bedded in sand and the load applied to the entire length of the top,

```
8-inch pipe broke when the weight per foot of length was 1363 to 2256 pounds
12   "     "     "     "     "     "     "     "     "     "   1227 to 2756    "
15   "     "     "     "     "     "     "     "     "     "   1261 to 2297    "
18   "     "     "     "     "     "     "     "     "     "   1464 to 2093    "
```

From similar tests made in 1897 F. A. Barbour of Boston deduced the expression $p = c\dfrac{t^{1.65}}{d}$, in which p is the pressure per lineal foot in pounds at the first cracking, t is the thickness in inches, d is the diameter in inches, and $c = 33,000$.

Tests made by T. H. Barnes on the strength of 12-inch vitrified clay pipe when acting as a beam between supports 2 feet apart gave the following results:

Thickness.	Cracked at (Pounds)	Broke at (Pounds)	Equal to (Lbs. per Lin. Ft.)	Remarks.
1″	1100	2750	1880	Fire-crack
1″	2000	2000	1330	
1″	2690	2810	1870	
1″	2220	2450	1630	
1″	2110	2535	1690	

The exact amount of pressure brought to bear upon a sewer by back-filling is uncertain. For a few feet of depth it probably bears the entire weight of the earth immediately above it. With granular material the proportion of pressure to weight of back-filling probably decreases but little, while with other soils it decreases more or less rapidly after the depth equals the width of the trench. But it is probable that, while the latter material gives an almost vertical pressure, the former acts more as a fluid, pressing normally to the surface of the sewer, and is not so liable to crush it. Little, however, is known on this point. From certain experiments in which natural conditions were only partially reproduced it was thought probable that for trenches 10 feet or more deep the percentage of weight of back-filling transmitted to the sewer equalled 1 — coefficient of friction of the material; that gravel transmits 36 per cent and wet clay 65 per cent of its weight; that up to 10 feet the percentage transmitted decreased from 100 per cent as the square or cube of the depth. If the depth of covering is small there is danger that outside weight from road-rollers or even heavy wagons may crush it. But this danger appears to be very slight when the depth of covering equals or somewhat exceeds double the width of trench.

The joints of vitrified clay pipe sewers are generally made of the bell-and-spigot pattern, as shown in Plate VIII, Fig. 8. The ring-joint (Fig. 9) is not now very extensively used, as its supposed advantages are found to be largely imaginary, while its disadvantages are not. It is almost impossible to make tight joints with the ordinary ring-joint and the expense is greater.

The joint of a hub-and-spigot pipe is made sometimes of clay, but in this country cement mortar is almost universally used. Clay has cheapness alone to recommend it as compared with cement. Other materials have been used for sewer-pipe joints, such as the Stanford preparation, a tar-and-sulphur compound. In Germany asphalt has recently been used and good results reported. Most of these materials are more expensive and less durable than Portland cement, and are probably to be preferred to it only under certain circumstances, if at all.

A glazed clay pipe offers a poor surface for cement to adhere to, and consequently with it an absolutely tight joint is almost impossible of construction. After a short period of use, however, a well-made joint of good cement will become so stopped with matter strained from out-filtering sewage as to be practically water-tight. But if the head of ground-water is greater than that of sewage the flow will be inward and the joint will probably not become tighter than it was at construction. Tighter joints could probably be made if the glazing were omitted or removed from the surfaces in contact with the cement.

If much sewage leaks out through a joint there is danger that the remaining fluid will not be sufficient to keep the sewer clean of deposits. But, as just stated, such a condition seldom continues for a long time after the sewer is put into use if the joints were well made.

Several modifications of the ordinary joint have been designed to overcome this difficulty, such as roughening the outside of the spigot end and the inside of the bell. One style of patent joint is shown in Plate VIII, Fig. 11. Such complicated joints are expensive and difficult both to manufacture and to lay, and are seldom used. If there is considerable ground-water it is better to lay the pipe as shown in Plate VII, Fig. 3, or to use light-weight or second-quality cast iron, or wrought iron or steel. Carefully made concrete or brick sewers may also be used for the larger sizes, of extra thickness to resist percolation.

The amount of ground-water which may leak through a cement joint depends very largely upon the shape of the bell and the manner in which the joint is made. If the annular cement-space in the bell is too small the cement is likely to be improperly compacted therein or not to enter at all at some points. Experiments seem to show that the deeper the ring of cement in the joint the less the leakage. If for any reason the cement draws away from either bell or spigot a leak is caused. Hence it seems best, particularly in wet soils, to use extra deep and wide sockets. The present standard of width is $\frac{3}{8}$ inch for pipe from 2 to 10 inches diameter and $\frac{1}{2}$ inch from 12 to 24 inches diameter, which, if always secured, should be sufficient. The depth of joint it would be well to have at least $1\frac{1}{2}$ inches greater than the thickness of the pipe; 2 inches would probably be better.

With poor joints the amount of leakage may be limited only by the amount of ground-water, but with the best of cement joints in very wet ground the leakage may amount to 5000 to 20,000 gallons per day per mile of sewer. In very many systems it is more than ten times this amount.

Experiment seems to show that neat Portland cement makes the tightest joints, Portland cement and sand 1 : 1 the next, natural cement and sand 1 : 1 the next, and natural cement neat the most porous joint.

Since the joint is the weak place in a pipe, the fewer joints there are the better. The expense of laying, also, is decreased by decreasing the number of joints. For these reasons the use of 3-foot rather than 2-foot lengths of pipe is advised. Vitrified clay pipes more than 3 feet long have not as yet been manufactured with success, but 3-foot lengths can be furnished by most pipe-manufacturers at the same price per foot as the 2-foot lengths. Some prefer to use the 2-foot lengths when the diameter of the pipe exceeds 15 or 18 inches, as the 3-foot lengths of the larger pipe would require a derrick for handling. Thirty-inch pipe is generally made $2\frac{1}{2}$ feet long.

There are some advocates and users of cement sewer-pipe, the most important in this country being the city (now borough) of Brooklyn, N. Y., which has used it almost exclusively for 35 years or more. It has the advantage over clay pipe that it can be moulded to exactly the size and shape desired, while the clay shrinks and sometimes warps in burning. It is therefore possible to obtain a sewer with a more uniform bore by using cement pipe; also to obtain the advantage (not very considerable under most circumstances) of a flat base, as shown in Plate VI, Fig. 1.

When this pipe is made of good cement and sand and this is properly proportioned and mixed it should give a material which will improve with age. It is, however, more difficult to detect the quality of a cement than of a vitrified clay pipe, and much worthless cement pipe has consequently been put upon the market. Clay pipe has a somewhat smoother surface, but this difference grows less with age, owing to the coating which forms on each.

Cement pipe weighs from 50 to 100 per cent more than clay pipe of the same diameter, and hence both freight and expense of handling are increased. Good cement pipe is in most places more expensive than good clay pipe.

ART. 47. MANHOLES, LAMP-HOLES, FLUSH-TANKS, ETC.

The purpose of manholes, as the name implies, is to give admittance to the sewers, which is necessary for the purpose of inspection and cleaning. They should therefore be sufficiently large to permit the passage through them of a man of average size.

Manholes are in general built immediately above a sewer and leading from it to the ground-surface. In the case of some large sewers in Europe they are built at one side of the sewer and connected with it by an underground passage, the chief advantages of which construction are the greater convenience for entering and the avoiding of manhole-heads in the street-paving. But this construction is very expensive and the passage is liable to be a collector of filth.

The size of vertical manholes is usually 24 inches, although sometimes only 22 or even 20 inches, diameter at the top, increasing towards the bottom to a size in which a man can work. The least size advisable for the bottom on lines of pipe sewers is 4 feet circular or 3 feet by 4 feet 6 inches oval. In manholes of this size the ordinary operations of inspection and cleaning of pipe sewers can be carried on. There is no particular advantage in having an ordinary manhole of more than 5 feet interior diameter.

Wherever possible the sides of the manhole should be built vertical from the side benches of the bottom (*ab* and *cd*, Plate VIII, Fig. 5) to a point 3 feet above, from which point they may be brought in with a straight batter to the smaller top, which is usually circular. Where the depth of the top of the sewer below the surface is less than 7 feet this construction becomes difficult, owing to the considerable angle which the upper walls must make with the vertical. The slope cannot well begin at a lower point than that stated and leave working-room at the bottom. If the depth of sewer is more than 5 feet this difficulty can be met by arching the walls (see Plate IX, Fig. 2), which construction requires careful workmanship. An alternative method, especially adapted to a depth of less than 5 feet, is to reduce the area of the manhole near the top by an offset, using either a brick arch or an iron beam to span the offset (see Plate IX, Fig. 3). If the manhole is more than 10 feet deep the diameter should increase more rapidly for the first 3 feet down from the top, being at least 2 feet 9 inches at that depth, as otherwise descent through the shaft will be difficult.

Descent through the manhole can be made by means of a ladder or a rope, but it is customary to build steps into the wall for this purpose. These may consist of protruding bricks or stones or cast- or wrought-iron pieces. The first offer but precarious footing, cast iron is not so reliable as

DETAIL PLANS.

Plate IX.

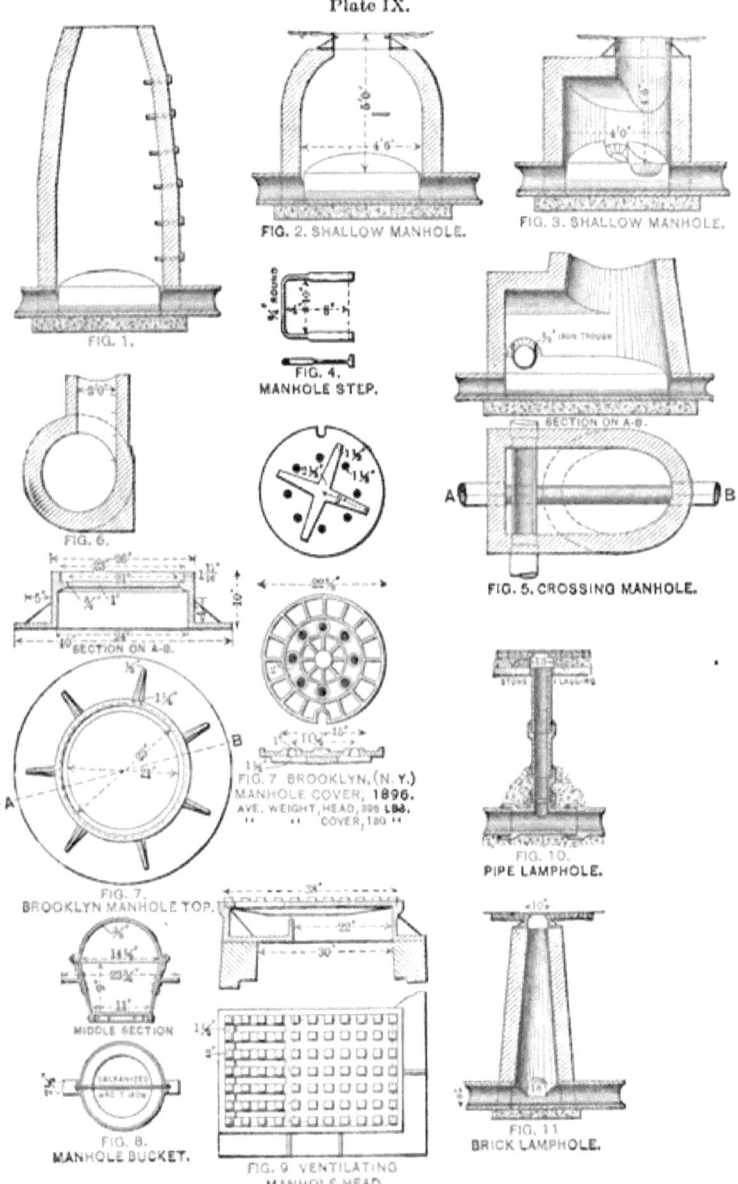

wrought and costs little, if any, less; the last is therefore recommended. These steps are made of various shapes. The simplest and probably as good as any is one made of a round bar bent as shown in Plate IX, Fig. 4. The steps should be placed about 15 inches apart vertically, and either directly under each other or alternating on each side of a vertical line, the former in narrow shafts.

Manholes oval at the bottom are well adapted to locations where there are no intersecting sewers; those circular, to points of intersection.

Where one sewer crosses another without intersecting it a manhole of special construction, permitting of inspecting each sewer, is desirable. Such a one is shown in Plate IX, Fig. 5, in which the upper sewer is continued through the manhole by an iron trough.

While at the junction of a pipe sewer-main and lateral the latter should be at a somewhat higher elevation than the former, the difference in elevation of the crowns of the two should not exceed 6 inches. To obtain this result the lateral may, if necessary, be lowered 2 or 3 feet at its end by increasing the grade from the previous manhole. If this would increase the depth of excavation by more than 3 or 4 feet a drop between the sewers may be made at the manhole. This should be so arranged that each sewer will be accessible for cleaning. The drop should not be made through the shaft of the manhole, but through a small smooth channel. A good design is that shown in Plate X, Fig. 8.

When sub-drains are laid under large sewers arrangements for cleaning them may be made as shown in Plate VI, Fig. 6, by a vertical branch opening into a manhole, or if they are under the centre of the sewer such a pipe may open into the sewer-invert, the opening being ordinarily tightly closed by a cap or plug. When the sub-drain is under a small sewer the branch pipe should lead into a manhole, opening either in the

DETAIL PLANS. 175

Plate X.

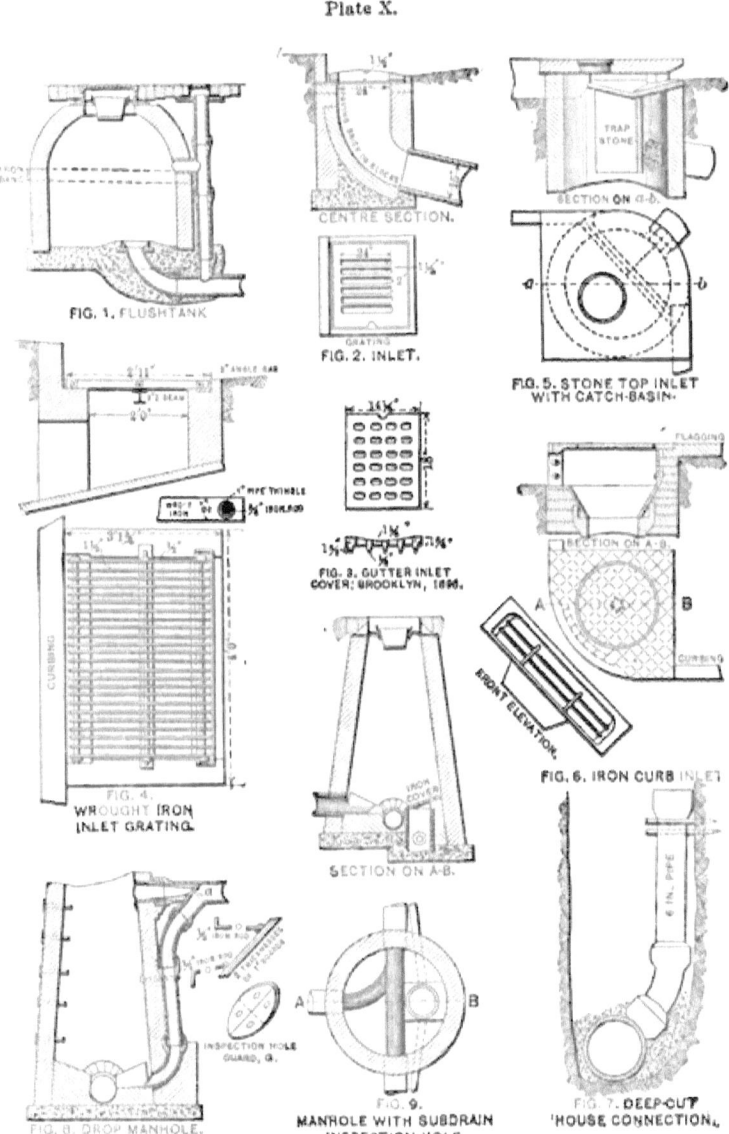

sewer-invert or, better, in the bench. In either case the opening should be plugged so that absolutely no sewage can enter it (see Plate X, Fig. 9).

Manholes of special design will be required by unusual conditions, but in all the three principal requirements of a manhole should be met: it should offer easy access to inspection and cleaning of the sewer, and ventilation of the same; it should also be so proportioned as to resist the pressure of the surrounding earth. For this last purpose the curved form is better than the polygonal.

Manholes for sewers larger than 30 to 36 inches are usually built up from the sewer-arch and have no special bottom construction. The sewer-invert under the manhole should be reinforced, however, if the ground is at all yielding. The manhole-shaft is sometimes placed on one side of the sewer both for strength and for facility of access (see Plate IX, Fig. 6).

The foundation of a manhole should be perfectly solid. If the soil is soft a plank platform may be used. Owing to the irregular shape of the bottom concrete usually gives better results as to strength, shape, and imperviousness than does brick-work. The bore of each sewer should be continued through the bottom by a smooth channel of uniform section and slope, either straight or with a continuous curve. This channel can be plastered with Portland cement, lined with brick or with split vitrified pipe. The last method gives the smoothest surface and is the one most likely to give a straight channel of uniform size. For curved channels, if split bends of the desired radius cannot be had, brick plastered with Portland cement is recommended. The channels should have vertical sides carried up to a point at least $\frac{2}{3}$ as high above the invert as the top of the sewer-pipe, and benches should slope up to the sides of the manhole at an angle of at least 10° or 15° with the horizontal.

The manhole walls are usually built of brick, 8 inches thick from the top to a point 10 or 12 feet below the surface, and increasing in thickness with the depth. If the bottom is a circle or a well-designed oval with no radius greater than 6 feet a 12-inch wall should be strong enough at any depth, unless the ground is a quicksand or similar material or is very wet. The outside of the manhole should be plastered with cement mortar to keep out ground-water or water used in settling the trenches, and to prevent the lifting of the top foot or two by freezing ground.

The top of the manhole is generally capped with an iron casting sufficiently deep to permit the laying close to it of brick or stone paving. This will be about 8 or 10 inches except where the paving is made for heavy or city traffic, where it may need to be 12 or 18 inches.

Whether the street is paved or not each manhole-head should be surrounded for a distance of at least 2 feet by stone or brick paving on concrete or sand foundation, the head being set $\frac{1}{4}$ to $\frac{1}{2}$ inch lower than the paving.

The cover should be sufficiently strong to support the heaviest wheel-pressure. It should be provided with ventilation-holes giving as much area of opening as possible. Its upper surface should be roughened to provide foothold for horses. The ventilation-holes should be through the elevated rather than through the depressed parts of the cover, since by this construction the stoppage of the holes by dirt and snow and the entrance of dirt into the sewer are considerably lessened. Such a manhole-head and -cover, as used in Brooklyn, N. Y., is shown in Plate IX, Fig. 7. Covers are sometimes provided with locks to prevent the opening of the manhole by unauthorized persons. Much trouble is in some instances caused by these locks, particularly in freezing weather. A better plan probably is to make the covers so heavy that they cannot readily be raised without the use of some strong implement adapted to this purpose.

More or less dirt will be sure to enter through the ventilation-holes and if allowed to reach the bottom of the manhole will tend, particularly in small sewers, to form stoppages. To prevent this a bucket of some kind should be suspended under the holes, smaller than the manhole-opening, that the air may pass up between the bucket and the walls, or a special construction of some kind should be designed for this purpose (see Plate IX, Figs. 8 and 9). These receptacles should be cleaned before they become filled with dirt, for which purpose the removable bucket of Fig. 8 is the more convenient. Another objection to Fig. 9 is the larger amount of street-surface occupied by the iron head.

Lamp-holes may be from 8 to 12 inches in diameter and are placed vertically above the sewer. They are sometimes made by placing in the pipe-line a T branch pointing upward and resting a vertical line of sewer-pipe in it. This is decidedly poor construction, as the branch pipe is liable to be crushed by the weight. The upright pipes should be supported by a foundation of brick or concrete or the entire shaft should be of brick. The latter is much to be preferred, since the pipe construction is almost sure to be pushed out of line by the settling of the back-filling.

The foundation of a lamp-hole should be firm, the invert formed as shown in Plate IX, Fig. 11. The head it would be well to provide with ventilation-holes, but this is seldom done.

A flush-tank should be tight. It should be so proportioned as to hold the required amount of water without increasing the head on the sewer beyond the limit set (Art. 26). The flush-tank is usually set at the upper end of a sewer-line, toward which much sewer-air rises, and the sewer should therefore be provided at that point with ample ventilation.

In spite of this many flush-tanks are so built as to afford the sewer absolutely no ventilation, forcing the adjacent houses to unwillingly, and usually unknowingly, provide it. Since flushing-siphons cannot permit of ventilation through their passages, a vent should be furnished the sewer just below the flush-tank. It is advisable to combine with this a lamp-hole, as in Plate X, Fig. 1. A still better plan is to place a ventilating-manhole just below, even in contact with, the flush-tank.

Flush-tanks are usually built of brick with concrete bottoms, the whole being made water-tight. Concrete or iron would probably be preferable in some cases.

The automatic flushing appliances in common use act on the principle of the siphon, the variations being in the method of starting the flow. Some have no moving parts whatever, such as the Rhoads-Williams and Miller tanks. Of those having moving parts the Van Vranken, which has a balanced tipping-pan at the foot of the siphon, is probably the best known. A number of other ideas have been used for flush-tanks, such as a tank on trunnions, which tips when full and returns to its original position when empty; a collapsing tube which, as the water rises in the tank, is extended upward by an attached float until it reaches its full length, when the water, still rising, overflows into and through it to the sewer, the tube meantime collapsing.

The outlet of the flush-tank should be at some elevation, the more the better, above the sewer. If no automatic appliance is used the opening of the flush-tank may be in the bottom, stopped by a plug or cap, which is raised by an attached chain when the tank is full; or it may be in the side and be opened and closed by a valve, either sliding or hinged.

If water is led to the flush-tank by a pipe this should be kept below the effect of frost, turning and rising to a higher level inside the flush-tank if necessary.

Inlets are made with and without catch-basins (see Art. 41), and the openings are sometimes vertical, sometimes horizontal, and sometimes inclined. Their purpose being to admit water from the roadway to the sewer, the opening of each inlet should be sufficiently large to admit all the water which can reach it from the heaviest rain whose run-off the sewer is designed to carry. It may be so designed that a smaller one leading to a house-sewer shall pass the water from small rains or the first washings of a rain, while another larger one leads to a storm-sewer. The opening should be at the gutter where the water flows, and which may be slightly depressed at this point. If horizontal in the bottom of the gutter one large opening is not permissible, but smaller ones, into which neither carriage-wheels nor feet of horses or pedestrians can enter, must be used. The plate through which these holes are made must be able to support the most heavily loaded wheels which are likely to come upon it. But this need not include exceptionally heavy loads, which usually keep to the centre of the street.

If the openings are through the face of the curb, in a plane either vertical or slightly inclined, they may be much larger. In some cases one large opening is used, entirely unprotected, through which children could and sometimes do fall. Except for this danger such a clear waterway is an excellent arrangement. But it is advisable to so place one or more bars across the opening as to remove the danger referred to.

The total area of opening required may be found approximately by the hydraulic formulas for flow through horizontal or vertical orifices or over weirs, as the case may be. In the case of openings less than 2 inches across in any direction an additional allowance should be made for the occasional stoppage of some of them by leaves, paper, etc. The vertical openings, being larger, are less liable to stoppage. If horizontal openings in the gutter are in the shape of slots they should run across the line of the gutter.

Between the openings and the sewer the channel should be straight or have as easy bends as possible, that the run-off may have an uninterrupted flow. The use of a catch-basin greatly interferes with this, the water seething and whirling in it during storms; consequently the channel connecting it with the sewer should be larger than if a simple inlet were used. In some instances a pipe leads directly from the opening to the sewer, either with or without a water-seal trap. It is better, however, to obtain a more substantial structure by setting under the opening a small basin with a curved bottom from which the pipe leads directly to the sewer. Where the opening is horizontal the basin is desirable to support the weight which may come upon the grating and, where a trap is used, to enable it to be placed below danger of freezing. It also facilitates inspection and cleaning of the connection-pipe (see Plate X, Fig. 2). Figs. 3 and 4 show two designs for inlet-gratings, the latter particularly adapted to admitting large quantities of water.

A catch-basin usually consists of a well under the inlet-opening and below the connection-pipe to catch the heavier matters. It is sometimes placed between the inlet and the sewer on the line of the connection-pipe, and sometimes at the sewer in connection with a manhole. To be at all efficient it should extend more than 18 inches below the connection-pipe, since a heavy rain will keep the water in it so stirred up as to wash out any deposits above that point. The bottom of the catch-basin should be covered with a flag-stone or the most substantial of concrete- or brick-work.

Inlet and catch-basin wells may be built of concrete or of stone, but are usually of brick. Catch-basin wells should be water-tight, that water may constantly cover the contents and lessen their odors. The gratings of catch-basins should be removable or the basins should be provided with manhole-openings and the wells be sufficiently large to be entered for the inspection and cleaning of the connection-pipes.

When the inlet-opening is vertical the well is usually under the curbing or sidewalk, and access to it is through a manhole-opening in the sidewalk. There is a great variety of inlet-tops for such construction, both cast iron and stone being used. The latter, where not too expensive, is usually preferable, being neater, more durable, and usually more like the contiguous sidewalk material than cast iron. A stone-topped inlet is shown in Plate X, Fig. 5, an iron-topped one in Fig. 6.

Traps are frequently placed in catch-basins or the connecting-pipes to prevent the exit of sewer-air, unwisely the author thinks (see Art. 41). The outside trap is usually a running or P pipe trap. Many varieties of inside trap have been designed, both fixed and movable. The former should not prevent access to the connection-pipe and hence should be at least 15 inches from its opening. Traps with movable parts should be as simple as possible in construction and compel the outflowing water to make the least possible number of angular changes of direction.

Instead of placing a catch-basin at each inlet it is sometimes preferable to place silt-basins along the line of the sewer at intervals of 1000 feet or more, with a manhole over each for ventilation and cleaning. These are particularly applicable to flat grades of storm sewers in the separate system. They consist of an enlargement of the sewer, and a depression of a foot or more in its invert, into which the heavier silt is washed, and from which it can be removed more easily than when deposited along a stretch of sewer. These, however, should not be used to encourage deposit, but only when deposits would occur along the sewer if they were not provided. Their advantage over inlet catch-basins is that the odors reach the outer air further from pedestrians, and that the difficulty and cost of cleaning is not so great. They should be used in sewers which carry house-sewage in exceptional cases only. Inlet catch-basins are generally preferable on lines of combined sewers where much heavy dirt reaches the inlet, or on storm-sewers where such dirt is washed in in very large quantities.

ART. 48. INTERCEPTORS AND OVERFLOWS.

The best form of interceptor to be employed is determined largely by the character of the system at the point of interception. If the house-sewage is to be intercepted from tributary sewers which originally discharged into a near body of water, the interceptor shown in Plate XI, Fig. 1, may be used. This "leaping weir," it is believed, was first used by Baldwin Latham about 1876. The exact length of opening required in the invert can be only approximately determined. It may be made smaller than is thought necessary and cut to the right size, which is ascertained by trial, after the sewer is in use. It will also probably be desirable to increase the length from time to time as the amount of house-sewage increases. The principal objection to this form of interceptor is that, although the storm-water may leap the opening, much of the sand and other heavy matter carried along the invert of the combined sewer will fall into the small intercepting sewer and be deposited there.

An interceptor which meets this objection, but which may more properly be called a divertor, is shown in Plate XI, Fig. 2.* The flap-valve shown is closed by the rising of the float, which occurs when the amount of sewage becomes greater than it is desired that the house-sewer carry. The joints of the mechanism should be of bronze. A sewer does not offer the best conditions for the continued proper working

---

* See *Engineering Record*, vol. XXXII, p. 41.

184 SEWERAGE.

Plate XI.

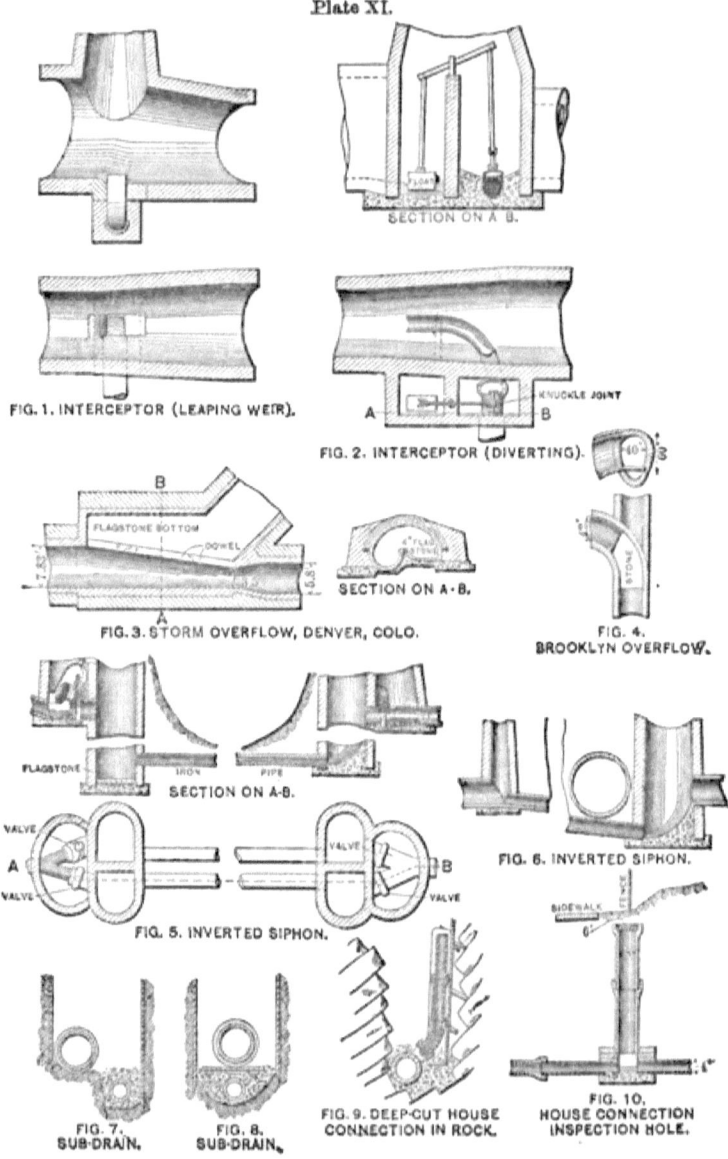

FIG. 1. INTERCEPTOR (LEAPING WEIR).
FIG. 2. INTERCEPTOR (DIVERTING).
FIG. 3. STORM OVERFLOW, DENVER, COLO.
FIG. 4. BROOKLYN OVERFLOW.
FIG. 5. INVERTED SIPHON.
FIG. 6. INVERTED SIPHON.
FIG. 7. SUB-DRAIN.
FIG. 8. SUB-DRAIN.
FIG. 9. DEEP-CUT HOUSE CONNECTION IN ROCK.
FIG. 10. HOUSE CONNECTION INSPECTION HOLE.

of any mechanism therein, but one so simple as this should give little trouble in its maintenance.

When a sewer, because of improper designing or of changed conditions, becomes too small to carry all the sewage coming to it, the excess above its capacity may be diverted to and carried by a relief-sewer or -sewers. A relief-sewer may cross under and receive the excess from several gorged sewers, or a single sewer may overflow into several relief-sewers placed at intervals along its length and leading to near-by outlets.

An outlet sewer-main to combined sewers is sometimes provided with overflow outlets at several points to avoid increasing the size of the main beyond the smallest necessary dimension, which is usually that which will carry sufficient storm-water to afford such dilution to the house-sewage as will render it unobjectionable to discharge this into an adjacent stream. The diversion into such a relief-sewer or relief outlet is ordinarily made by means of an overflow, constructed as shown in Plate XI, Fig. 3, or as in Fig. 4, where the relief-sewer was constructed after the smaller sewers had long been in use.

Art. 49. Inverted Siphons; Sub-drains; Foundations.

Inverted siphons are usually circular in section, since always flowing full; usually of metal, since always under pressure, although the metal may be lined with brick or other material. The size required has already been referred to. When laid under water they should be so weighted or covered with earth or stone as to prevent their floating when pumped empty for inspection or cleaning, and should be absolutely tight. The inverted siphon is made sometimes to slope from both ends to a point near mid-length, sometimes with a vertical drop at one end, sometimes at both ends. The first should be adopted only when the siphon is sufficiently large to permit the entrance of a man. When not of such a size it should be straight from end to end. This will usually require a shaft at one, sometimes at each, end, which may also serve as a manhole. It is in most cases advisable to place a catch-basin at the foot of such a shaft, although in place of this a basin in the bottom of an enlargement of the sewer just above the siphon is sometimes employed. A siphon with catch-basins is shown in Plate XI, Fig. 5, the valves on the ends of each siphon-pipe permitting either siphon to be closed to sewage and pumped out for inspection, while the other is in use.

Unless a siphon under water is of large size and in tunnel or laid in a trench in a rocky bottom it should be protected from undermining by currents, or movement by shifting bottoms or channels. This protection is usually afforded by driving a row of sheet-piling on each side of the pipe, the space between these being in most cases excavated and filled with concrete. The softer the material in the bottom and the stronger the currents the deeper the sheeting should be driven. If the bottom is too hard to permit of driving sheeting, large stone rip-rap may be placed on both sides and over the siphon.

A sewer must sometimes pass either under or over an obstruction—such as a water-main, another sewer, etc.—by a siphon, either inverted or erect. The latter requires greater care in construction and constant attention to maintain a vacuum at the summit, and the former is in the majority of cases the preferable construction. Such a siphon is usually a few feet in length only and under but little head. A manhole should be placed over or near it when the sewer is 24 inches or more in diameter, since it will probably need more frequent cleaning than the other parts of the line. If the sewer is less than 24 inches diameter a manhole should be placed at the upper end of the siphon (which should be straight from end to end), and at the lower end also, although a lamp-hole may be substituted here if the siphon is not over 150 feet long, and makes only an angle and not a vertical rise at this point. For such a case see Plate XI, Fig. 6.

Sub-drains are placed either directly beneath the sewer or at one side of the trench. When there are no artificial foundations under the sewer the latter position is to be preferred, but is in some instances much more difficult and expensive, particularly in quicksand. The sub-drain should be surrounded with broken stone or clean gravel, varying preferably from the size of a hickory-nut to that of a pea. There should be at least 3 inches of this under the drain and 6 inches at its sides and top. In quicksand or similar material these dimensions should be increased 50 to 100 per cent. This stone should be well compacted to prevent future settlement. The joints of the drain should be slightly open and a 5- or 6-inch strip of cheese-cloth or burlap wrapped around the pipe at the joint to keep out the dirt. Or, if bell-and-spigot pipe is used, a piece of jute may be calked loosely into the joint for this purpose.

If a sewer were laid directly over this there would be danger of a settlement of the same and of leakage resulting. For this reason the sub-drain should be laid at one side of the trench when the soil is firm, as in Plate XI, Fig. 7. In quick or running sand this is practically impossible unless the trench is very wide or unless close sheathing be driven on each side of the sub-trench and carried below its bottom; such sheathing not to be removed after the sub-drain is laid. It would usually be better and cheaper than this to lay the sub-drain in the centre of the trench (which must of course be close-sheathed in quicksand), and on the stone filling, when levelled off, to place a continuous platform on which to lay the sewer. Such construction is shown in Plate XI, Fig. 8. A still better construction in any but firm soils is to lay a pipe sewer in concrete, as in Plate VII, Fig. 3. Where a foundation is necessary for the sewer the sub-drain construction is easily arranged. See Plate VI, Fig. 6, and Plate VII, Fig. 10.

The sub-drain should be laid to grade as carefully as the sewer itself. It is seldom that a sub-drain can be so arranged that inspection can be made of it, and therefore perfectly straight alignment is not necessary; but there should be no sharp angles in its line, which might cause obstructions or interfere with the future cleaning of it. If cellars and basements are to be connected with this drain Y branches should be inserted to permit of such connections, and should be covered similarly to the house-sewer branches.

When house or combined sewers are placed with their tops more than 4 or 5 feet lower than the average cellar depth in that locality it is advisable to place a standing house-connection above each branch, bringing it to within 3 to 5 feet of the average depth of the cellar bottoms, but stopping at least 7 or 8 feet from the surface. This is to avoid compelling each householder along the line to dig down to a deep sewer branch in order to make a connection. These standing connections are built while the sewer-trench is open, and are covered at the top with a cap or cover similar to house-branches. They should not merely rest in the branch, but a foundation of concrete or brick masonry should support each. The vertical pipes should be held in place during back-filling, as by stakes driven into the bank. In the case of a rock cut, or where the banks are not firm, the standing connection may be inclosed by a vertical trough of planks, between which and the pipe earth is packed, this trough being held firmly in place until the trench is filled and tamped. If the banks are liable to cave, sheathing should be driven at each such connection, and neither it nor the braces removed when the trench is filled. A standing house-connection in firm soil is shown in Plate X, Fig. 7. One in a rock cut is shown in Plate XI, Fig. 9.

A sewer in soft soil, like any other structure, requires a foundation. Since the weight is not comparatively great the service of the foundation is more often to distribute the pressure and prevent local settling or heaving than to prevent the subsidence of the sewer as a whole. This purpose is usually achieved by use of a cradle (Plate VI, Fig. 4) or a platform of plank (Plate VI, Fig. 5), the former in comparatively firm soils like damp sand or loam, the latter in swamp-muck, quicksand, etc. Where muck or other soft, water-sogged soil is encountered it may be necessary to drive piles and rest a timber platform upon these. Such a foundation is shown in Plate VI, Figs. 3 and 6. Where a platform is used it is necessary to fill the sub-invert spaces of the sewer with masonry. All sewers in soft soils should have their inverts arching downward to resist the upward thrust of the ground between the side walls, since the weight of the masonry is largely concentrated in these walls.

In rock excavations no part of the pipe sewer should come within 6 inches of the rock bottom, and the space between this and the sewer should be filled with sand or gravel thoroughly tamped to prevent settlements of the invert; or the pipes should be bedded in concrete, in which case the rock may be taken out only to the under side of the pipe. If the sewers are built of masonry this should be carried to rock everywhere under the invert.

CHAPTER IX.

SPECIFICATIONS, CONTRACT, ESTIMATE OF COST.

ART. 50. DEFINITION AND CLASSIFICATION OF SPECIFICATIONS.

PUBLIC work is frequently, if not in the majority of cases, done by contract by a " party of the second part " who is paid for this work by the city, the " party of the first part." That the contractor shall do the work as the city desires it is necessary that he be instructed what is desired and that he bind himself to follow the instructions. This should all be recorded in writing for the protection of both the city and the contractor. The agreement to perform the work on the one hand and to pay for the same on the other is called a contract and is generally accompanied by a bond under which the contractor places himself to perform the work as directed.

The directions, called "specifications," "consist of a series of specific provisions, each one of which defines and fixes some one element of the contract. These clauses relate, in general: first, to the work to be done; second, to the business relations of the two parties to the contract." (Johnson's " Engineering Contracts and Specifications.") The clauses in specifications for sewer construction referring to the work to be done may be classified as those: first, defining the character of the material to be employed; second, giving directions, dimensions, etc., for excavating and back-filling; third, setting forth the methods to be employed in the construction of the sewer-barrel and appurtenances, including foundations; fourth, stating the requirements of the completed work, tests to be made, etc.; fifth, giving general directions for the conduct and maintenance of the work, employment of labor, etc. Disposal plants will require separate specifications, varying with the character of the disposal employed. No general form for such can be given. Other special features of a system will call for special clauses.

The clauses relating to the business relations of the two parties to the contract may be classified as relating to: first, time of commencement and of completion and rate of progress of the work; second, character of labor and appliances to be employed; third, measurement of and payments for the work; fourth, contractor's protection of and responsibility for lives and property; fifth, abandonment, cancellation, assignment of contract, etc.; sixth, definition of names and terms employed.

The specifications are generally accompanied by a set of plans which form a part of the specifications and contract. These together should set forth the work to be done so clearly as to leave no point for future dispute. Care should be taken that contradictory instructions are not given, but that all parts of both plans and specifications mutually agree. Too great profuseness should be avoided as confusing to contractor, inspector, and engineer. Many engineers insert provisions which they have no intention of enforcing under ordinary conditions, merely to be on the safe side, or which aim at theoretic perfection of details which cannot be attained in practice (of which fact their inexperience may make them ignorant). The fact that some clauses in a specification cannot be enforced is apt to detract from the effectiveness of the others. It is better to make only such requirements as experience shows are desirable and practicable and give the contractor to understand that these will be rigidly enforced.

No foresight can predict all the emergencies which may arise in sewer construction. To provide for these it must be agreed that the engineer can modify plans or methods of work during construction, as well as increase or decrease quantities. Work not at first specifically provided for may be made the subject of separate contract, or if but small in quantity may be done under the original contract as extra work, to be paid for at its cost plus such a percentage for profit (generally 10 or 15 per cent) as is fixed in the contract.

ART. 51. SPECIFICATIONS FOR MATERIALS.

A set of specifications for sewer construction will be given and discussed in the succeeding pages. Some alterations and additions will probably be required to adapt them to any particular case, but it is thought that they will be of considerable service as an illustration of both matter and form. Clauses in brackets are given as alternatives, the one preferred by the author being placed first; the same also holding true with reference to the lettered paragraphs.

Paragraph 1. a. Sewer-pipe.—All pipe and specials, unless otherwise specified, shall be of the best quality, salt-glazed, vitrified clay sewer-pipe of the hub-and-spigot pattern; both body and bell shall [have a thickness not less than $\frac{1}{12}$ the inside diameter of the pipe] [be of standard thickness]. Each hub shall be of sufficient diameter to receive, to its full depth, the spigot end of the next following pipe or special without any chipping whatever of either, and also leave a space of not less than one half inch all around for the cement-mortar joint; it shall also have a depth from its face to the shoulder of the pipe on which it is moulded at least 2 inches greater than the thickness of said pipe. Straight and curved pipe having diameters up to and including 15 inches shall be furnished in 3-foot lengths. Branches may be in 2-foot lengths. All pipe and specials shall be sound and well burned, with a clear ring, well glazed and smooth on the inside and free from broken blisters, lumps, or flakes which are thicker than $\frac{1}{8}$ the nominal thickness of the pipe and whose largest diameters are greater than $\frac{1}{8}$ the inner diameter of said pipe; and pipe and specials having broken blisters, lumps, and flakes of any size shall be rejected unless the pipe can be so laid as to bring all of these defects in the top half of the sewer. No pipe having unbroken blisters more than $\frac{1}{4}$ inch high shall be used unless these blisters can be placed in the top of the sewer. Pipes or specials having fire-checks or cracks of any kind extending through the thickness shall be rejected.

No pipe shall be used which, designed to be straight, varies from a straight line more than $\frac{1}{8}$ inch per foot of length; nor shall there be a variation between any two diameters of a pipe greater than $\frac{1}{24}$ the nominal diameter.

No pipe shall be used which has a piece broken from the spigot end deeper than $1\frac{1}{2}$ inches or longer at any point than $\frac{1}{2}$ the diameter of the pipe; nor which has a piece broken from the bell end if the fracture extends into the body of the pipe, or if its greatest length is greater than $\frac{1}{2}$ the diameter of the pipe, or if such fracture cannot be placed at the top of the sewer. Any pipe or special which betrays in any manner a want of thorough vitrification or fusion or the use of improper or insufficient materials or methods in its manufacture shall be rejected.

(*Many engineers specify a depth of bell only 1 inch* "greater than the thickness of said pipe," *but it is difficult to make tight joints in actual practice with such bells. Frequently the defects of sewer-pipe are not referred to in detail, but the acceptance or rejection made optional with the engineer or inspector.*

If cement pipe is used the following paragraph may be substituted for 1. a.)

Paragraph 1. b. Sewer-pipe.—All pipe and specials, unless otherwise specified, shall be of the best quality of cement sewer-pipe, of the [hub-and-spigot] [bevelled-joint] pattern; it shall have a thickness not less than $\frac{5}{8}$ inch plus $\frac{1}{12}$ the diameter of the pipe. [Each hub shall be of sufficient diameter to receive, to its full depth, the spigot end of the next following pipe or special, without any chipping whatever of either, and also leave a space of not less than $\frac{1}{2}$ inch all around for the cement-mortar joint; it shall also have a depth from its face to the shoulder of the pipe on which it is moulded at least 1 inch greater than the thickness of said pipe.] [The bevel on each pipe shall be at least 25 per cent longer than the thickness of said pipe, with an even and firm edge.]

All pipe shall be in 3-foot lengths and in section shall truly correspond to their nominal shapes. Each pipe shall have a flat base making exact right angles with the vertical axis of the pipe and with a width equal to $\frac{2}{3}$ the interior horizontal diameter of said pipe. The inside surface of the pipe shall be smooth and true, and no pipe shall be patched with cement or otherwise. Any pipe will be rejected which is not of fine, sound, and dense material throughout, or which shows the use of poor materials or imperfect mixing or compacting.

Paragraph 2. a. Drain-pipe.—Pipe for sub-drains shall be of vitrified clay sewer-pipe in 1- or 2-foot lengths [of the hub-and-spigot pattern] [without bells or sleeves]. It shall comply with the specifications for sewer-pipe in so far as these refer to thickness, quality, and vitrification of material, blisters, lumps, flakes, cracks, and breaks; except that the engineer may at his option accept pipe having small fire-cracks or checks.

Paragraph 2. b. Drain-pipe.—Pipe for sub-drains shall be composed of the best quality of drain-tile of [circular] [horse-shoe] cross-section in [one-] [two-] foot lengths. They shall be hard-burned and without cracks or any considerable departure from their nominal shape, size, or cross-section.

Paragraph 3. Brick.—For all brick-work none but the best quality of sound, hard-burned, perfect-shaped bricks, presenting a regular and smooth surface, shall be used. After being thoroughly dried and immersed in water for 24 hours they shall not absorb more than 10 per cent by weight of water. Shale brick, if used, shall be composed of rock thoroughly ground and shall be homogeneous throughout and uniformly burned.

Paragraph 4. Paving-stone.—This shall consist of hard granite or trap-rock, uniform in grain and texture. The blocks must be rectangular in form, not less than 3 nor more than 4 inches in either length or breadth, nor less than 4 nor more than 5 inches in depth, and so split and dressed with true surfaces that on neither top, ends, nor sides shall there be a projection from the general surface exceeding $\frac{1}{4}$ inch.

(*This stone is used for inverts where there is excessive velocity in the sewer or impact from falling water.*)

Paragraph 5. Masonry-stone.—Stone for foundations and backing shall be of a sound and durable quality, free from cracks and seams, having top and bottom beds approximately parallel. No stone shall be less than 4 inches thick, 12 inches long, and 8 inches wide.

Paragraph 6. Iron Castings.—All iron castings shall be made from a superior quality of gray iron, remelted in the cupola or air-furnace, tough and of even grain, and shall possess a tensile strength of not less than 18,000 pounds per square inch. Test-bars of the metal 3 inches by $\frac{1}{2}$ inch, when placed upon supports 18 inches apart and loaded in the centre, shall have a transverse breaking load of not less than 1000 pounds, and shall have a total deflection of not less than ⅜ inch before breaking. These test-bars shall be poured from the ladle at any time the engineer directs, before or after the castings have been or while they are being poured. All castings shall conform to the shape and dimensions shown upon the drawings and shall be clean and perfect, without blow- or sand-holes or defects of any kind. No plugging or other stopping of holes will be allowed. The castings shall be thoroughly cleaned of all lumps and subjected to careful hammer tests, after which they are to be dipped in a bath of coal-tar pitch heated to at least 200° Fahr.

Iron pipe shall comply with the above specifications, except that the engineer may, at his option, receive a pipe having a limited number of small sand- or blow-holes on its exterior surface. No portion of the shell of the pipe shall have a less thickness than ——— (*this thickness can generally be made the least which will permit of handling of the pipe without danger of breaking it, and non-uniformity of shell is not objectionable if payment is not made by weight*).

Paragraph 7. Wrought Iron.—All wrought iron must be tough, ductile, and fibrous, of a uniform quality, free from crystalline structure, cinders, flaws, or cracks. In bars it must have an ultimate strength of 50,000 pounds per square inch. Iron which has been burnt in the forge will be rejected. Each wrought-iron piece furnished shall correspond in all respects to the dimensions specified.

Paragraph 8. Sand.—All sand shall be clean, sharp, and free from loam, clay, or vegetable matter. It shall not be so fine that each grain on the surface of a pile cannot be readily noted with the naked eye, nor shall it be exceedingly coarse when used for brick masonry.

(*Dirt in sand can usually be detected by rubbing a small*

amount on the palm, which will be soiled by any clay or loam present.)

Paragraph 9. Cement.—Unless otherwise specified all cement shall be of the best quality of natural cement, and when tested neat in briquettes (Am. Soc. C. E. standard) shall show a tensile strength of at least 75 pounds after 1 hour in air and 23 hours in water and of at least 150 pounds after 1 day in air and 6 days in water. Cement for brick masonry or pipe-joints, when these are laid in wet ground, shall be quick-setting and show a tensile strength of at least 100 pounds per square inch after 24 hours. Pats of neat cement made on glass and brought to a thin edge shall show no checks after setting in boiling water.

When specified Portland cement shall be used. This shall show a tensile strength of at least 400 pounds per square inch in a 7-day test made as above, and pats of the same shall show no checking. The cement mixed neat and stiff into pats $\frac{1}{2}$ inch thick shall develop "initial" set in not less than 20 minutes and "hard" set in not less than 45 minutes after mixing, except in the case of quick-setting cement to be used as specified above.

The engineer shall be allowed to test all cement and notice of its receipt by the contractor must be made to the engineer at least 48 hours in advance of its use upon the work. Any cement not satisfactory to him shall be at once removed from the work.

Paragraph 10. Packing.—Packing may consist of flax, jute, oakum, or hemp, clean and with long fibres loosely twisted into strands.

Paragraph 11. Timber.—All timber and planking used in cradles, platforms, and foundations shall be of spruce, or timber equally as good, straight, sound, free from sap, shakes, large, loose, or decayed knots, worm-holes, or other imperfections which may impair its strength or durability. Piles shall be of sound, straight, live spruce or yellow-pine timber, of lengths specified by the engineer for each locality. They shall be not less than 6 inches in diameter at the smaller end. The bark shall be removed in all cases.

ART. 52. EXCAVATION.

Paragraph 12. Classification of Materials.—All materials excavated shall be classified as either earth or rock. No material shall be classified as rock which cannot be removed more cheaply by drilling and blasting than by picking, except that any boulder measuring $\frac{1}{2}$ cubic yard or more shall be so classified, whether blasted or removed bodily; but such boulder shall not be returned to the trench without being first broken up.

Paragraph 13. Excavation of Trench.—The trench shall be excavated along the line designated by the engineer and to the depth necessary for laying the sewer or sub-drain at the grade given by him. In the case of pipe sewers it shall be 1 foot wider at the bottom than the outside diameter of the pipe, and for brick sewers as wide as the greatest external horizontal width of the structure to be placed therein, without any undercutting of the banks. Where, in the opinion of the engineer, the original earth is sufficiently compact and solid for the foundation of the work the contractor shall excavate the bottom of the trench to conform to the external form and dimensions of the invert or foundation as ordered. For pipe sewers the bottom of the trench under each bell shall be so hollowed out as to allow the body of the pipe to have a bearing throughout on the trench bottom and permit of making the joint. In case a trench be excavated at any place, excepting at joints, below the proper grade it shall be refilled to grade with sand or loam thoroughly rammed, without extra compensation unless the extra excavation was ordered by the engineer.

The material excavated shall be laid compactly on the side of the trench and kept trimmed up so as to be of as little inconvenience as possible to the travelling public and to adjoining tenants. Where the street is paved the paving shall be kept separate from the other material excavated. (*It is generally desirable to place the paving material on the side of the trench which is to be left open for travel, and the earth upon the other.*) All streets shall be kept open for travel and the engineer reserves the right to require the use of **excavating-machinery** if necessary to insure this.

No tunnelling will be allowed except by written permit, with restrictions, from the engineer. When tunnelling, the contractor will **excavate the material to such cross-section as may be designated**, using timbering or other tunnel-lining and shoring satisfactory to the engineer. The location and size of any shafts, and the location of pumps, derricks, boilers, and other machinery, must be approved by the engineer (*see Art. 69*). The engineer shall have the right to limit the amount of trench which shall be opened or partly opened at any one time in advance of the completed sewer, and also the amount of trench left unfilled.

The contractor shall not, without permission from the engineer, remove from the line of the work any sand, gravel, or earth excavated therefrom which may be suitable for refilling the trench until the same shall have been refilled.

Paragraph 14. **Pumping and Bailing**.—The contractor shall furnish all necessary machinery for the work, shall pump, bail, or otherwise remove any water which may be found or shall accumulate in the trenches, and shall perform all work necessary to keep them clear of water while the foundations and the masonry are being constructed or the sewer laid. In no case, unless by special permission of the engineer, shall water be allowed to run over the invert or foundation or through the sewer until the cement is satisfactorily hardened. The disposal of the water after removal shall be satisfactory to the engineer.

Paragraph 15. Shoring and Sheathing. — Whenever necessary the sides of the trench shall be braced and rendered secure and either open or close sheathing used, to the satisfaction of the engineer; such sheathing and bracing to be left in until the trench is refilled, all such bracing and sheathing being done at the contractor's expense. Sheathing left in permanently by the order of the engineer, and only such, will be paid for at the price bid. When left in the trench sheathing shall be cut off at a point about 1 foot below the surface. The contractor shall, at his own expense, shore up and otherwise protect any building which may, in the opinion of the engineer, be endangered by the work.

Paragraph 16. Railway-crossings. — When any railway-lines are to be crossed or interfered with specific directions as to the time and manner of doing this work will be given by the engineer, and the contractor shall conform to such directions. He shall be allowed for material furnished and made part of the permanent construction, so far as it may be additional to that indicated on the plan, but all other work shall be done at his own cost.

Paragraph 17. Interference with Existing Structures and Watercourses. — In excavating and back-filling trenches and laying the sewer care must be taken not to move or injure any gas-, water-, sewer-, or other pipes, conduits, or structures without the order of the engineer. If necessary the contractor shall, at his own expense, sling, shore up, and secure, and maintain a continuous flow in said structures, and shall repair any damage done to them and keep them in repair until the final acceptance of the completed works, leaving them in as good condition as when uncovered. Should it be necessary to move the position of a pipe or conduit this shall be done in accordance with the instructions of the engineer, and the contractor shall be allowed for material furnished and made part of the permanent construction, so far as it may be additional to that indicated upon the plans, and for labor performed on such additional construction, but all other work shall be done at his own expense.

At such street-crossings and other points as may be directed by the engineer the trenches shall be bridged in a secure manner, so as to prevent any serious interruption of travel upon the roadway and sidewalks and also to afford necessary access to public and private premises. The material used and mode of constructing such bridges and the approaches thereto must be satisfactory to the engineer; the cost of all such work must be included in the regular price bid for the sewer. (*Crossings should not be tunnelled under, since it is almost impossible to so refill the tunnels as to prevent after-settlement, but should be bridged. Direct access to the street should be given to fire-engine houses and usually to livery-stables.*) All fire-hydrants shall be left uncovered and accessible. The contractor shall at his own expense provide for all watercourses, gutters, and drains interrupted by the work, and replace them in as good condition as he found them.

Paragraph 18. Rock Trenches.—When the excavation for a pipe sewer or drain is made through rock or other material too hard to be readily or conveniently removed for admitting the hubs of the pipe the trench shall be excavated at least 4 inches deeper than the grade of the outside bottom of the pipe and [filled with concrete up to and around such pipe, as shown upon the plans] [refilled to such grade with sand or loam, free from stones or other hard substances, thoroughly rammed]. When rock is encountered in the trench it shall be stripped of earth and the engineer notified and given proper time to measure the same before blasting.

All rock removed which has not been measured by the engineer will not be estimated as rock excavation. Measurement for rock excavation will be limited to 6 inches on either side of the sewer, and trench-slopes of 8 vertical to 1 horizontal. In all cases of blasting the blast shall be carefully covered with heavy timbers chained together, and the engineer may limit the number of simultaneous discharges. Not more than 30 pounds of dynamite shall be kept on hand at one time in any one place. No blasting shall be done within 40 feet of the finished sewer or 10 feet of an uncovered gas- or water-pipe, and the end of the finished sewer shall be covered or stopped with plank or earth during each blast. (*If the sewer-end is not so protected there is a possibility of stones flying into the sewer and also of the concussion of air opening the joints.*)

ART. 53. CONSTRUCTION.

Paragraph 19. Foundations. — When timber or pile foundations other than those shown in the plans are necessary, in the opinion of the engineer, special designs will be furnished the contractor, who, in accordance with such designs, shall place such foundations in position satisfactory to the engineer. Planking in platforms shall be laid in the manner directed, closely joined, and each plank spiked to each cap or sill with nails or spikes of a length at least $2\frac{1}{2}$ times the thickness of the plank. If cradles or platforms are laid directly upon the ground this must be graded perfectly even and smooth to receive them and give a good and firm bearing throughout. If caps or sills are used the spaces between them and under the planking must be filled with good earth thoroughly rammed.

Where piles are used they shall be driven to refusal, unless extending more than 10 feet below the foundation, when they shall show a penetration in inches under the final blow not greater than $\frac{wh}{L} - 1$, in which L is the weight to be borne by each pile, w is the weight of the hammer in pounds, and h its fall in feet. After driving, the piles shall be sawed off truly and evenly at the proper elevation for receiving the caps, which shall be fastened to them with 1-inch drift-bolts of a length twice the depth of the sill, holes for such bolts having first been bored with a ⅞-inch bit. If any pile shall be out of line more than ⅛ the diameter of its upper end the engineer may refuse to estimate it and may order another driven in its place.

Concrete or stone-masonry foundations shall be constructed where ordered in a manner similar to that specified for " Concrete " and " Stone Masonry."

Paragraph 20. Concrete. — Concrete, unless otherwise specified, shall be composed of 1 part by bulk of natural cement, 3 parts of sand, and 5 parts of broken stone, gravel, or furnace-slag of approved quality, free from dust and dirt and broken so as to pass in every way through a 2-inch ring. All material shall be actually measured for each batch, the cement compacted in barrels as received (or, if received in bags, an equivalent quantity as ascertained by trial), the sand and stone in similar barrels or specially prepared boxes. In mixing, the sand shall be spread out upon a suitable platform or box and the cement deposited upon this; these shall then be thoroughly mixed dry until the whole is of an even, uniform color, when sufficient clean water shall be added to form a thick paste. The stone, which has previously been thoroughly wet, shall then be added and the whole shall be quickly and thoroughly mixed, until every stone is coated with mortar, water being gradually added by sprinkling, if necessary, to obtain a better consistency. If mixing be done by machinery it shall produce a mixture equally as good as by the above method. Concrete must not be mixed in quantities greater than required for immediate use, and any which has begun to set shall not be retempered or used in any way. Concrete shall be deposited in layers not to exceed 9 inches in thickness, and settled by thorough light ramming, sufficient to bring water to the surface. One course shall follow another as rapidly as possible. Where fresh concrete is to be placed in contact with that already set or partly set all loose stone or concrete not thoroughly compacted shall be removed from the surface of the latter, which shall be washed clean of all dirt and given a thin coat of mortar. If such a surface be hard set it shall previously be thoroughly water-soaked. When concrete is in place all wheeling, working, or walking on it must be prevented until it is firmly set, and until such time it shall be kept damp and protected from the sun.

Such forms and centres as may be necessary to give the finished concrete the desired form shall be furnished by the contractor without extra charge. These shall be sufficiently stiff and substantial to retain the concrete firmly in place, and shall not be withdrawn until the same has set to the satisfaction of the engineer. No concrete shall be made or used when the temperature is below 35° Fahr. without the permission of the engineer, whose instructions and restrictions for such use shall be followed. (*When an entire sewer is composed of concrete a better quality, generally made of Portland cement, is used for the invert, and the inside is plastered.*)

Paragraph 21. **Stone and Brick Masonry.**—Stone and brick masonry, unless otherwise specified, shall be laid with mortar composed of 1 part by measure of natural cement to 2 of sand, mixed as specified for concrete mortar. No mortar shall be used after it has set or partially set.

Stone masonry must be laid true and by line and built of the exact dimensions shown in the plans of the work. All stones shall be laid upon their natural beds and roughly squared on the joints, beds, and faces, the stone breaking joints at least 6 inches, and with at least one header for every three stretches. Headers shall be at least 3 feet long or extend entirely through the wall. No stone once bedded shall be lifted by spalling, but any spalls used must be embedded in the mortar before setting the stone. Each stone shall be floated to place in a full bed of mortar and every joint thoroughly filled with the same. No dressing of stone upon the wall will be allowed. (*For river- or retaining-walls further specifications should be added as to thickness of joints, character of face dressing, etc.*)

For brick masonry in straight walls or sewers none but whole, sound brick shall be used. For manholes, flush-tanks, and similar work a limited number of half brick may be used, not to exceed $\frac{1}{8}$ of the whole in any case. Unless the engineer direct otherwise each brick shall be thoroughly wetted immediately before being laid. (*If the brick absorbs practically no water this wetting should be omitted, as likely to cause the brick to slide on the mortar and cause uneven work.*) It shall be laid with a full, close joint of cement mortar on its bed, ends, and side at one operation. In no case is mortar to be slushed in afterward. Special care shall be taken to make the face of the brick-work smooth, and all joints on the interior of a sewer shall be carefully struck with the point of a trowel or pointed to the satisfaction of the engineer. Where pipe-connections enter a sewer or manhole " bull's-eyes " shall be constructed by laying rowlock courses of brick around them, the cost of such construction being included in the regular price bid for the sewer or appurtenance. Around pipe more than 15 inches in diameter 2 rowlock courses shall be laid.

Brick-work in sewers shall be laid by line, each course perfectly straight and parallel to the axis of the sewer. Joints appearing in the sewer shall in no case exceed $\frac{1}{4}$ inch in width. Sewers shall conform accurately in section and dimensions to the plans of the same. All inverts and bottom curves shall be worked from templets accurately set, the arches are to be formed upon strong centres accurately and solidly set, and the crowns keyed in full joints of mortar. No centres shall be drawn until the arch masonry has set to the satisfaction of the engineer and refilling progressed up to the crown. They shall be drawn with care, so as not to crack or injure the work. The extrados is to be neatly plastered with cement mortar $\frac{1}{2}$ inch thick, the arches being cleaned and wetted just before plastering. The end of each section of brick sewer shall be toothed or racked back, and before beginning the succeeding section all loose brick at the end shall be removed and the toothing cleaned of mortar. All brick-work shall be thoroughly bonded, adjacent courses breaking joints at least $\frac{1}{3}$ the exposed length of the brick.

Stone blocks shall be laid in Portland-cement mortar composed of equal parts by measure of cement and sand. Joints shall not exceed $\frac{3}{8}$ inch in width. The face of the masonry shall be such that there shall be no projection beyond the general surface exceeding $\frac{1}{4}$ inch. All joints shall be cleaned out to a depth of $\frac{1}{2}$ inch and pointed with neat Portland-cement mortar. All stone-block work shall be laid in other respects as specified for brick-work.

If there should be any distortion of the sewer before acceptance this shall be corrected by tearing down and rebuilding. No local patching will be allowed, but when repairs are necessary a section shall be removed at least 3 feet long and including the entire arch, or the entire sewer if the defect is in the invert. Leakage of ground-water into the sewer shall be similarly corrected, unless it can be prevented by calking the joints with oakum saturated in cement, with wooden plugs, or other material acceptable to the engineer.

Paragraph 22. Laying Pipe Sewers.—Previous to being lowered into the trench each pipe shall be carefully inspected, and those not meeting the foregoing specifications shall be rejected, and either destroyed or removed from the work within 10 hours; except that pipe suitable for sub-drains may be used for that purpose, but shall be kept apart from the sewer-pipe. All lumps or excrescences on the ends of each pipe shall be removed before it is lowered into the trench. No pipe shall be laid except in the presence of the engineer or his authorized inspector, and the engineer may order the removal and relaying of any pipe not so laid. The trench shall be excavated in accordance with Paragraphs 13 and 18. No sewers shall be laid within 10 feet of the excavating or 40 feet of the blasting. Pipes having any defects which do not cause their rejection shall be so laid as to bring these in the top half of the sewer, and if the bell or spigot be broken the defective place must be liberally covered with neat-cement mortar, reinforced with a piece of pipe or pipe-ring if the engineer so direct.

The pipes and specials shall be so laid in the trench that after the sewer is completed the interior surface thereof shall conform accurately to the grades and alignment fixed and given by the engineer. All adjustment to line and grade of pipes laid directly upon the bottom must be done by scraping away or filling in the earth under the body of the pipe, and not by blocking or wedging up. Before laying, the interior of the bell shall be carefully wiped smooth and clean, and the annular space shall be free from dirt, stones, or water. [(*For hub-and-spigot joints.*) A narrow gasket of packing dipped in cement grout shall be properly calked into each joint, after which cement mortar shall be introduced therein. Such gasket shall be in one piece, of sufficient length to reach entirely around the pipe and of a thickness sufficient to bring the bottoms of the two pipes to the same level. No joint shall be cemented until the gaskets of the next two joints in advance are properly inserted. Special care must be taken to properly fill with mortar the annular space at the bottom and sides as well as at the top of the joints. After such space has been filled, the cement having been compacted with a wooden or iron calking-tool, a neat finish shall be given to the joint by the further application of similar mortar to the face of the hub so as to form a continuous and even bevelled surface from the exterior of said hub to the exterior of the spigot all around.] [(*For bevelled joints.*) The bevels shall each be covered with a layer of cement at least $\frac{1}{4}$ inch thick and the spigot pipe steadily pushed home with some force. A band of cement at least $\frac{1}{4}$ inch thick and 3 inches wide shall then be neatly wiped around the outside of the sewer at the joint.] All water must be kept out of the bell-hole during laying, or else such bell-hole must be completely filled out with the cement mortar specified or with concrete, for which mortar or concrete no extra compensation will be allowed. The interior of the joint shall be wiped clean of cement by a wad made of a sack filled with hay, large enough to tightly fill the pipe and attached to a rod or cord, which shall at all times be kept in the sewer and pulled ahead past each joint as soon as it is cemented. The mortar used shall be composed of [1 part cement to 1 of sand] [neat cement] wet to a thick paste. (*Engineers do not agree as to the advisability of using neat-cement mortar. Experiment seems to show that natural cement gives a tighter joint if mixed with sand, Portland if used neat.*) Natural cement shall be used under ordinary conditions, but the engineer may require the use of quick-setting or of Portland cement when he thinks it necessary.

As soon as the cementing of any joint has been completed the bell-hole under the hub must be carefully and compactly filled with sand, loam, or fine earth, so as to hold the external mortar finish of said joint securely in its place. Refilling shall also be made with selected material, free from stones, carried halfway up the sides or circumference of the entire length of pipe and compacted with a proper tamping-tool. The trench shall then be filled to a point at least 2 feet above the top of the pipe with material containing no stone larger than 2 inches in any dimension.

While the pipes and specials are being laid in each section between manholes or other permanent openings light from the remote end of the section shall remain constantly in plain view throughout the entire length of such section or division. Sections between openings will in general not exceed 300 to 400 feet; in particular cases the distance may be somewhat greater.

At such places as will be directed by the engineer, branches will be inserted in the sewer for future connections. Each branch thus inserted shall be closed by a thin vitrified stoneware cover or plug, which shall be placed before the special pipe is lowered into the trench. The covers shall be so inserted and cemented in as to prevent any water entering the sewer, at any time before their removal, through such branches. The entire cost of furnishing and setting such covers shall be included in the regular price bid for branches. Where directed by the engineer deep cut connections (Fig. - - -) shall be constructed as shown upon the plans.

Any omission of the required branches, manholes, lampholes, or other special constructions indicated upon the plans, or that may be specially ordered beforehand by the engineer, shall be corrected by the contractor at his own expense.

Before leaving the work for the night or at any other time the end of the sewer shall be securely closed with a tightfitting plug.

Paragraph 23. Laying Sub-drains.—Sub-drains shall be laid in sub-trenches excavated of the dimensions and in the location shown upon the plans, and of such depth as is necessary to lay the pipe at the grade given by the engineer. This sub-trench shall be filled with clean broken stone or gravel, not less than ¼ inch nor more than 1 inch in any dimension, up to the drain invert; this broken stone or gravel being laid in by hand or shovels and lightly compacted, so that there may be no future settlement. On this the drain-pipe shall be laid accurately to grade, having first been inspected and all pipe not meeting the specifications having been rejected. A piece of cheese-cloth or similar material satisfactory to the engineer, at least 5 inches wide and twice as long as the outside circumference of the pipe, shall be laid on the broken stone with its centre under the joint between two pipes; first one end and then the other of this shall be carried over the pipes and under the opposite side, care being taken to keep the cloth spread out and its centre over the joint. The pipes shall be separated by a space of about ¼ inch. The space between the pipes and the sides of the sub-trench shall then be carefully filled with broken stone or gravel as specified above, carefully compacted, which material shall be similarly placed to a depth of 6 inches above the pipe. Where directed by the engineer this stone-filling shall be covered by hemlock plank, to be paid for as "timber in foundations." If any earth or other material shall fall into the sub-trench while the laying of stone filling is proceeding, such material and the adjacent stone-filling shall be removed and clean stone be put in its place.

Where directed by the engineer branches shall be inserted in the sub-drain for future connections. These shall be closed as specified for sewer branches, and the specification as to the omission of sewer specials shall apply to sub-drain specials also.

Paragraph 24. Regular Appurtenances.—Manholes of the various kinds—line, intersection, drop, etc.—lamp-holes, flush-tanks, inlets, and other appurtenances shall be built where the engineer may direct, in size, form, thickness, and all other respects in accordance with the plans, but manholes whose height exceeds 12 feet shall have walls 12 inches thick below that depth. All appurtenances shall be brought up accurately to the grade given by the engineer. Great care shall be taken to make the channels in manholes and lampholes conform accurately to the sewer grade. In the case of pipe-sewers split pipe shall be used for the inverts to these channels where possible. Where a curve in the channel or some other condition prevents this the channel shall be formed of bricks on edge, set in Portland-cement mortar. Brick channels shall be lined with neat Portland-cement mortar $\frac{1}{4}$ inch thick, and the inverts shall be exactly semi-circular of the diameter of the pipes which they connect. If these be of different diameters the channel shall taper uniformly from one size to the other.

Flush-tanks and inlets shall be plastered on the outside with $\frac{1}{2}$ inch of cement mortar; and on the inside shall be given three coats of thin Portland-cement grout, without sand, applied with a brush, each coat being allowed to set before the next is applied. (*This will be more certain to make a water-tight construction than plastering with* mortar.)

Care shall be taken to place the inlet tops, when these are in the sidewalk, exactly in line with the curb, and to place the bottoms of the openings or the gratings exactly on the gutter grade given.

All manholes and flush-tanks shall be fitted with steps similar to those shown on the plans, and spaced 15 inches apart vertically. All tops or other fittings shall be set during the construction or at the completion of each appurtenance, in a firm, neat, and workmanlike manner.

All concrete, stone, or brick masonry shall conform to the specifications given in Paragraphs 20 and 21. Each appurtenance shall be begun within 24 hours of the time it is reached in the laying of the sewer, and shall be completed and the excavation closed as expeditiously as possible.

ART. 54. BACK-FILLING AND CLEANING UP.

Paragraph 25. a. Back-filling.—In back-filling sewer-trenches loose, fine earth, free from stones, shall be used up to a point 2 feet above the sewer, and shall be thoroughly compacted in 6-inch layers with hand-rammers. The remainder of the trench shall contain not more than $\frac{1}{8}$ broken rock, and no stone of this shall weigh more than 50 pounds. If necessary to meet this requirement the contractor shall supply suitable material for back-filling. The filling of the trench above the level of 2 feet above the sewer shall be rammed in 9-inch layers, or, when directed by the engineer, the trenches shall be water-tamped. Water-tamping shall be done in each case as directed by the engineer. All back-filling shall be done by hand and in no case shall scrapers or ploughs be used. In back-filling of tunnels or under railroad tracks especial care shall be taken to thoroughly compact the material. (*The question of back-filling is a very troublesome one. In most soils, when the diameter of the sewer does not exceed one sixth of the depth of the trench, all the earth excavated can be returned without leaving any ridge and without any appreciable after-settlement. But this can be done only at considerable expense—from 4 to 12 cents for each cubic yard of back-filling—by careful ramming or water-tamping; in tough clay no way has yet been found to accomplish this. When the trench is through fields or unpaved streets this extra payment is not generally warranted by the benefits derived; but through paved streets it generally is. The above specifications are similar to those ordinarily used, but contractors generally understand that they will not be enforced except in well-paved streets, and bid accordingly. It is preferable to leave the option* *confessedly with the engineer as to whether the trench shall be tamped, and pay for the tamping which is ordered, having it well done. The following specification is offered as a substitute, to be rigidly enforced.*)

Paragraph 25. b. Back-filling.—In back-filling sewer-trenches loose, fine earth, free from stones, shall be used up to a point 2 feet above the sewer, and shall be thoroughly compacted in 4-inch layers by hand-rammers, there being two rammers to each shoveller. Rammers for this purpose shall weigh from 4 to 6 pounds each, and have not to exceed 10 square inches of face. The remainder of the trench shall contain not more than one third broken rock, and no stone of this shall weigh more than 50 pounds. If necessary to meet this requirement the contractor shall supply suitable material for back-filling. Unless otherwise specified the trench above the level of 2 feet above the sewer shall be filled by hand with this material up to within 1 foot of the surface, and the remainder of the filling shall be made of fine material containing no stone having any dimension greater than 2 inches. The filling shall be crowned above the trench, having a height above the street surface of twice as many inches as the top width of the trench in feet, and neatly rounded off, the paving material previously removed, if any, being spread evenly over the top. After refilling, and for 6 months after the completion of this contract, the contractor shall from time to time refill any settlements which may occur, constantly maintaining the trench in a neat and safe condition, and deliver it over in that condition at the end of that time. Hand-ramming or water-tamping shall be used where directed, and as follows, an additional sum being paid therefor according to the price bid.

For hand-ramming the earth shall be spread by shovels in 4-inch horizontal layers and solidly compacted with rammers weighing from 6 to 8 pounds and having a face of not to exceed 20 square inches. There shall be two rammers to every shoveller, and the former shall be of at least as great strength and efficiency as the latter. The paving shall be restored in as good condition as found, being given a crown of $\frac{1}{2}$ inch over the trench, but not, in the case of macadam or gravel paving, overlapping the old paving. During back-filling no sheathing which is to be drawn shall at any time extend into earth which is being rammed, but it shall be drawn so as to be always above it, if it cannot be at once entirely removed.

For water-tamping the earth shall be levelled off in horizontal layers 2 feet thick and flooded with water until, after standing for 5 minutes, water shall just show on the surface, when another layer shall be thrown in and flooded. This shall be continued up to within 2 feet of the surface and allowed to stand for a few hours. The last 2 feet shall then be put in and hand-rammed as specified above, and the paving relaid.

No water shall be turned into the trench until all cement-work in sewers and appurtenances shall have had full time to set.

If a trench is rammed or water-tamped any earth which may have slipped or caved from the bank shall be thrown out of the trench and the space refilled and tamped in the same way as the trench proper, without extra compensation.

Paragraph 26. Street Surfaces.—In all paved, macadamized, or improved streets generally the surface of the trenches shall be finished without needless delay, in the most workmanlike manner, with the same kind of roadway improvement that was removed in excavating the trench, and so that the underlying courses, as well as the finished surface, shall conform to the remainder of the roadway, and shall in every respect be equal in quality, character, materials, and workmanship to the street improvement existing over the line of the trench immediately previous to making the excavation. The expense of restoring the pavement or improvement must be included in the price per lineal foot of sewer.

Paragraph 27. Cleaning up.—As soon as the trench has been refilled and paving replaced all stones, plank, or other refuse material of whatever description deposited and left by the contractor on the streets shall be removed therefrom and the said streets restored in all respects to the same condition as before the trenching was commenced. All surplus earth which may be left on the street after the trenches have been refilled as specified above shall be regarded as the property of the contractor, and must be removed as soon as possible at his expense.

Paragraph 28. Final Inspection.—Upon notification by the contractor of the completion of the work herein contracted for the engineer will carefully inspect all sewers, appurtenances, and all other work done by the contractor. In each stretch of pipe sewer intended to be straight light shall be visible from one end to the other. Any broken or cracked pipe shall be replaced with sound ones. The interior of brick sewers shall be of the required shape and dimensions, sound and of a uniform surface. Any deposits found in the sewers, protruding cement or packing, shall be removed and the sewer-bore left clean and free through its entire length. There shall be no appreciable amount of leakage into any stretch of sewer. All underdrains shall discharge water freely and give evidence of having a clean and open bore. All manholes, lamp-holes, and other appurtenances shall be of the specified size and form and of a neat appearance, and their tops shall be set to the proper grade. In general the work shall comply with these specifications, and if found not to do so in any respect shall be brought to the proper condition by cleaning, pointing, or, if necessary, excavating, and rebuilding, all at the expense of the contractor. But if it be found after uncovering any pipe or other work at the order of the engineer that no defect exists, or that the defect was not due to any fault of the contractor, then the expense of this shall be borne by the city.

ART. 55. GENERAL PROVISIONS, PAYMENTS, ETC.

Paragraph 29. General Provisions.—If any alterations in plan directed by the engineer diminish the quantity of work to be done they shall not constitute a claim for damages nor for anticipated profits, and any increase or decrease shall be paid for or deducted according to the quantity actually done, and at the price established for such work under this contract.

The work shall be prosecuted in such manner and from as many different points, at such times and with such force as the engineer may, from time to time during the progress of the work, determine.

The contractor will be furnished with a set of drawings showing the details and dimensions necessary to carry out the work, dimensions in figures thereon having precedence over the scale. These plans and a copy of these specifications are to be kept constantly at the work by the contractor or his authorized foreman. The plans submitted to contractors for proposals are to be interpreted in conjunction with the specifications, and descriptions of the character of the work appearing on the plans are made a part of these specifications. No deviations from the drawings will be allowed without the direction of the engineer to that effect.

Should it be necessary at any time to move monument-stones or other permanent records the contractor shall not disturb them until given permission by the engineer.

The contractor shall provide suitable stakes, plank, and forms, and render such assistance to the engineer, at his own expense, as may be necessary to establish lines and grades for the guidance of his work, and shall carefully preserve said points at all times.

If any person employed by the contractor on this work shall appear to be incompetent or disorderly he shall be discharged immediately on the requisition of the engineer, and such person shall not again be employed on the work.

Paragraph 30. Responsibility for Injuries.—The contractor shall be responsible for injuries to person and property inflicted during the prosecution of the work, and for all damages caused by the negligence of the contractor or any of his employees, workmen, or servants, and the city may at its discretion withhold the amount of such injury or damage from any estimate due him which may be needed to make good such damages or injuries, and the city shall not in any wise be liable therefor.

The contractor shall place sufficient lights on or near the work and keep them burning from twilight to sunrise, shall erect suitable railing or protection about the open trenches, and provide all necessary watchmen on the work by day or night, for the safety of the public.

Paragraph 31. Imperfect Work.—When any work or material is found to be imperfect, whether passed upon or not by the inspector, the said work shall be taken up and replaced by new work at any time prior to final acceptance.

If the contractor shall be notified by the engineer of any requirements or precautions neglected or omitted, or of any work improperly constructed, he shall at once remedy the same, and if he fail so to do the engineer, under the direction of the city, shall perform such work at the contractor's expense and deduct the same from amounts due or to become due the contractor.

Paragraph 32. Unnecessary Delays.—In case of any unnecessary delay, in the opinion of the engineer, he shall notify the contractor in writing to that effect. If the contractor should not, within 5 days thereafter, take such measures as will, in the judgment of the engineer, insure the satisfactory completion of the work the engineer may then, under authority from the city, notify the aforesaid contractor to discontinue all work under this contract, and it is hereby agreed that the contractor is to immediately respect said notice and stop work and cease to have any rights to possession of the ground. The engineer shall thereupon have the power to place such and so many persons as he may deem advisable, by contract or otherwise, to work at and complete the work herein described, and to use such materials as he shall find upon the line of said work, or to procure other materials for the completion of the same, and to charge the expense of said labor and materials to the aforesaid contractor; and the expense so charged shall be deducted and paid by the party of the first part out of such money as may be then due, or at any time thereafter become due, to said contractor under and by virtue of this agreement or any part thereof; and in case such expense is less than the sum which would have been payable under this contract if the same had been completed by the party of the second part [he] [they] shall be entitled to receive the difference, and in case such expense is greater the party of the second part shall pay the amount of such excess so due.

Paragraph 33. Extra Work.—If any work of the general nature of the work herein contracted for, but for doing which a bid has not been especially made, shall need to be done the contractor shall do the same under the direction of the engineer, and shall receive therefor the actual cost of labor and material used plus ten per cent (10%) for superintendence and use of tools, but he shall not be entitled to receive payment for any work as extra work unless ordered by the engineer to do the same as such. No claim for extra work will be allowed if not made before the payment of the next following monthly estimate.

Paragraph 34. Time of Commencement and Completion.
—The party of the second part agrees to begin the work herein contracted for within two weeks of the awarding of the contract, and to fully complete the work herein specified on or before the ...... day after the awarding of said contract, but the party of the first part may extend the time of completion should they deem it for the best interest of the city. [It is expressly understood that the party of the second part agrees to pay all expenses, such as engineering and inspection, that the city may be put to by reason of the work being incompleted at the time specified in the contract.] [For each day after the time specified that the contract remains uncompleted $25 will be deducted from the amount due the contractor, and for each day by which the contract is completed previous to the time specified the contractor shall be entitled to a bonus of $25.] [For each day after the time specified that the contract remains uncompleted $25 will be deducted from the amount due the contractor, and it is hereby expressly understood that said sum shall be deemed and taken in all courts to be the liquidated damages for the non-performance of the work in the manner aforesaid, and not in the nature of a penalty.]

Paragraph 35. Definitions.—Whenever the word "engineer" is used in the specifications it refers to the engineer in charge of the work and also to his authorized agents.

The "party of the first part" is the city by and for which the work herein described and referred to is being done, and the "party of the second part" is the person or persons contracting to do said work.

The word "sewer" in its general sense in these specifications refers to the sewer-barrel and to any bends, slants, branches, or other details joined to or forming a part thereof.

The word "appurtenance" refers to all manholes, lamp-holes, flush-tanks, inlets, and all structures forming a part of the sewerage system, but not included in the term "sewer."

Paragraph 36. Position of the Engineer.—The engineer shall have the final decision on all matters of dispute involving the character of the work, the compensation to be made therefor, or any question arising under this contract. He shall, as representing the city, have the option of making any changes in the line, grade, plan, form, position, dimensions, or material of the work herein contemplated, either before or after construction is begun, and all explanations or directions necessary for carrying out and completing satisfactorily the different descriptions of work contemplated and provided for under this contract will be given by said engineer.

Paragraph 37. Duties of the Contractor.—The contractor must perform the work contracted for strictly according to these specifications, and follow at all times, without delay, all orders and instructions of the engineer in the prosecution and completion of the work and every part thereof, and constantly be on the ground or be represented by a duly qualified person to look after the work and receive instructions.

Paragraph 38. Measurements and Payments.—Measurements of sewers and drains shall be taken from the centre of the uppermost manhole or flush-tank on each line to the centre of the manhole at its junction with a main or lateral, or to the centre line of such main or lateral at the junction, including all branches, manholes, or other appurtenances along the line. The depth by which sewer prices will be graded will be measured from the surface of the ground to the under side of the sewer-pipe or masonry or of the timber platforms or foundation-sills. The price bid for sewers or drains shall include furnishing all material and labor for excavating, shoring, constructing the sewer or drain in accordance with the specifications and plans, back-filling, restoring the street-surface as previously specified, and for all matters in connection therewith heretofore specified as being so included. Measurements of connections shall be taken from the outside (bell) end of the branch to the upper end of the connection-pipe. Branches shall be paid for by the piece at the price bid, which shall include the cost of furnishing and fixing plugs in said branches where necessary.

Deep-cut connections shall be paid for at the prices bid for " deep-cut connections," " pipe," " concrete," and " timber in foundations," according to the actual quantities used, the bid for " deep-cut connections " including the combining of these and the setting of, and extra care in back-filling around, the pipe.

Flush-tanks shall be paid for at the price bid for each particular size of tank, this to include the tank complete as set forth in the drawings and specifications, including the excavation and back-filling, ventilation-pipe and iron head.

Ordinary manholes and lamp-holes shall be paid for on the basis of a depth of 8 feet, with an additional amount for each foot by which the depth exceeds 8 feet, the price bid to include excavating and back-filling, furnishing and setting iron castings and steps, and completing the whole as set forth in the plans and specifications. The depth of flush-tanks, manholes, and lamp-holes shall be measured from the invert of a pipe sewer, or the springing of a brick sewer, to the top of the iron head when properly set.

The price bid for " crossing-" and " drop-manholes " shall be an additional sum over and above the bid for the same as a regular manhole, and shall be held to cover furnishing material for and constructing the crossing or drop device as shown in the plans, as an addition to the regular manhole. The bid for the crossing-manhole shall be a lump sum; that for the drop-manhole shall be per vertical foot, measuring from the invert of the lower to that of the upper sewer.

The price bid for inlets, catch-basins, and other appurtenances shall include the excavation and back-filling, and furnishing all materials and constructing each appurtenance in strict conformity to the plans and specifications.

The price bid for stone paving shall be per square foot, and shall be over and above the price for the sewer or manhole in which it is laid.

The price for stone, brick, or concrete masonry not otherwise provided for shall be per cubic yard by actual measurement in place, provided such dimensions do not exceed those indicated or implied in the plans or instructions of the engineer.

Iron-work, both cast and wrought, shall be paid for by the pound, but no payment for iron-work as such shall be made for the heads or steps or other devices included in the manholes and other appurtenances as shown in the plans and specifications. Cast-iron pipe will be paid for at the price per foot bid for the same.

The price for timber in foundations shall include the furnishing and setting of the same. The price bid for furnishing piles shall be for the lengths actually delivered, where these do not exceed those ordered by the engineer. The price for driving piles shall be per foot, measured from the bottom of the pile when driven to the surface of the ground in which it is driven, and shall include cutting off the piles at the elevation given by the engineer.

The price bid for tamping trenches shall be by the cubic yard of trench above a point 2 feet above the top of the sewer, the bottom width heretofore specified being allowed and side slopes of 1 in 15 in earth and 1 in 8 in rock.

The engineer on the first of each month, or within 5 days thereafter, during construction, will estimate approximately the amount of work completed during the preceding month, according to these specifications, and eighty-five per cent (85%) of the amount due beyond the reservations herein made will be paid the contractor on or before the 15th day of each month for the work of the preceding month.

When the contract shall have been completely performed on the part of the contractor the engineer shall proceed to make final measurements and estimates of the same, and shall certify the same to the city ...., who shall, except for cause herein specified, pay to the contractor, on or before the 15th day after such completion of the contract, the balance which shall be found due, excepting therefrom such sum as may be lawfully retained under any provision of this contract.

ART. 56. CONTRACT.

Accompanying the specifications and bound with them should be the contract proper, of which a form is given:

THIS AGREEMENT, made and concluded the .... day of .... in the year One Thousand Eight Hundred and ...., by and between the City of ...., of the first part, and ...., Contractor, of the second part,

WITNESSETH, That the said party of the second part (has) (have) agreed, and by these presents (does) (do) agree with the said party of the first part, for the considerations herein mentioned and contained, and under the penalty expressed in a bond bearing even date with these presents and hereto attached, to furnish at (his) (their) own proper cost and expense all the necessary material and labor, except as herein specially provided, and to excavate for and build, in a good, firm, and substantial manner, the sewers indicated on the plans now on file in the office of the city engineer, and the connections and appurtenances of every kind complete, of the dimensions, in the manner, and under the conditions herein specified; and (has) (have) further agreed that the engineer in charge of the work shall be and is hereby authorized to inspect or cause to be inspected the materials to be furnished and the work to be done under this agreement, and to see that the same correspond with the specifications.

The party of the second part hereby further agrees that (he) (they) will furnish the city with satisfactory evidence that all persons who have done work or furnished material under this agreement, and are entitled to a lien therefor under any law of the State of ...., have been fully paid or are no longer entitled to such lien, and in case such evidence be not furnished as aforesaid such amount as the party of the first part may consider necessary to meet the lawful claims of the persons aforesaid shall be retained from the moneys due the said party of the second part, under this agreement, until the liabilities aforesaid may be fully discharged and the evidence thereof furnished.

The said party of the second part further agrees that (he) (they) will execute a bond in a sum equal to 25 per cent of the contract price, secured by a responsible Indemnity or Guarantee Company of, or authorized by law to do business in, the State of ...., and satisfactory to the city, or by at least three responsible freeholders of .... County satisfactory to the city, for the faithful performance of this contract, conditioned to indemnify and save harmless the said city, its officers or agents, from all suits or actions of every name or description brought against any of them for or on account of any injuries or damages received or sustained by any party or parties, by or from the said party of the second part, (his) (their) servants or agents, in the construction of said work, or by or in consequence of any negligence in guarding the same or any improper materials used in its construction, or by or on account of any act or omission of the said party of the second part, or (his) (their) agents, in the performance of this agreement and for the faithful performance of this contract in all respects by the party of the second part; and the said party of the second part hereby further agrees that so much of the moneys due to (him) (them), under and by virtue of this agreement, as shall be considered necessary by the said city may be retained by the said party of the first part, until all such suits or claims for damages as aforesaid shall have been settled and evidence to that effect furnished to the satisfaction of said city.

The said party of the second part hereby agrees to receive the following prices as full compensation for furnishing all materials, labor, and tools used in building and constructing, excavating and back-filling, and in all respects completing the aforesaid work and appurtenances, in the manner and under the conditions before specified, and as full compensation for all loss or damages arising out of the nature of the work aforesaid, or from the action of the elements or from any unforeseen obstructions or difficulties which may be encountered in the prosecution of the same, and for all expenses incurred by or in consequence of the suspension or discontinuance of the said work, and for well and faithfully completing the same and the whole thereof according to the specifications and requirements of the engineer under them, to wit:

(Insert here spaces for making bids, being careful to include every item for which bids are invited. As an example:)

For all 36-inch brick sewer, trenches from 6 to 8 feet deep.............	$——	per lineal foot
For water-tamping..................	——	per cubic yard
For each manhole 8 feet deep, complete	——	
For each vertical foot of manhole more than 8 feet deep, 8-inch wall.......	——	

For each vertical foot of manhole more
than 8 feet deep, 12-inch wall......   ——
For timber foundations..............   —— per M B. M.

etc.   etc.

And the said party of the second part further agrees that (he) (they) will not assign, transfer, or sublet the aforesaid work, or any part thereof without the written consent of the city, and that any assignment, transfer, or subletting without the written consent aforesaid shall in every case be absolutely void.

IN WITNESS WHEREOF the said party of the second part (has) (have) hereunto set (his) (their) hand and seal and the said party of the first part has caused these presents to be sealed with its common seal and to be signed by the ....
.... on the day and year above written.

It is recommended that the engineer refer to Johnson's "Contracts and Specifications," where will be found a full discussion of the subject from both the legal and engineering standpoint.

ART. 57. ESTIMATE OF COST.

It is generally desirable, and frequently required by law, that a careful estimate be made of the cost of the work to be done. For this purpose map, plans, specifications, and profile should be carefully studied to obtain quantities, and the amount of rock to be excavated, quicksand, and ground-water ascertained, and in general as careful a study made of the conditions as a contractor would make before bidding. Also the prices of materials should be obtained, including the cost of getting them upon the ground, and from these as close an estimate made as possible of the actual cost of constructing the system. To this should be added 10 to 100 per cent for

SPECIFICATIONS, CONTRACT, ESTIMATE OF COST.

profit and contingencies, the latter amount when the work is to be done under great risks and subject to possible losses.

Out of a dozen bids made on one sewer contract there are generally one or two quite low, two or three others quite high, and the remainder more or less close together midway between these, and usually representing a fair price for the work, which also the engineer's estimate should do. The estimate should not be too low, as this often gives rise to suspicion of intentional deception, and if made the basis of an appropriation of funds for construction may lead to a forced curtailment of the amount of work done. On the other hand, an unduly high estimate may discourage any appropriation whatever. Probably no act of the sewerage engineer is more readily appreciated by the public at large than the making of an estimate closely approximating the actual cost.

The cost of brick, lumber, and sand varies with each locality and should be obtained from local dealers. That of cement and pipe varies little except with the freight, and this variation is slight between different places in the same section and on main freight-lines.

A schedule price has been adopted by all sewer-pipe makers, from which large discounts are allowed. Such a list is given below. The discount allowed contractors for car-load lots at present (1898) near New York City is about 80 to 82 per cent for sizes under 24 inches.

TABLE No. 17.

LIST PRICES OF VITRIFIED CLAY SEWER-PIPE, AND WEIGHTS OF STANDARD PIPE.

Size, inches	2	3	4	5	6	8	9	10	12	15	18	20	24
Straight pipe, per foot	0.14	0.16	0.22	0.25	0.30	0.45	0.55	0.65	0.85	1.25	1.70	2.25	3.25
Bends	0.40	0.50	0.65	0.85	1.10	1.80	2.25	2.75	3.50	4.75	6.50	7.50	11.00
Branches, 2 feet long, each	0.63	0.72	0.90	1.13	1.35	2.03	2.48	2.93	3.83	5.63	7.65	10.13	14.63
Traps, each	1.00	1.50	2.00	2.50	3.50	5.50	6.50	7.50	10.00	...	...	...	...
Weight of straight pipes per foot, pounds	6	7	10	12	16	22	26	32	45	63	84	98	130

Slants are charged 50 per cent more than plain pipe, and

measured on the long side of the slant, but none less than 12 inches long.

Each additional branch or trap is charged branch price.

Double-strength pipe is allowed 10 per cent less discount than standard pipe.

Increasers are pipe with the hub on the small end and reducers with the hub on the large end, and are charged double the price of 2 feet of pipe of the size of the large end.

Channel or split pipe, which is pipe cut in two or more pieces lengthwise, costs $\frac{3}{8}$ the price of whole pipe.

Stoppers or plugs for closing pipe cost $\frac{1}{3}$ as much as 1 foot of pipe of the size in which they are used.

TABLE No. 18.

LIST PRICES OF DRAIN-TILE.

Size, inches	2	2½	3	4	5	6	8	10	12
Price, straight pipe	.012	.015	.020	.030	.040	.055	.080	.140	.200

36- and 38-inch WOODEN-STAVE PIPE in Los Angeles, Cal., cost $2.25 to $2.50 per foot, complete.

Light-weight CAST-IRON PIPE, first quality, cost in 1898 about as follows:

TABLE No. 19.

COST OF LIGHT IRON PIPE.

Size, inches	4	6	8	10	12	14	16	18	20	22	24	27	30	33
Cost per foot	.18	.27	.35	.41	.49	.60	.75	1.00	1.25	1.40	1.62	1.82	2.20	2.50

One barrel of cement, used neat, should lay the following lengths of sewer, pipes 3 feet long:

TABLE No. 20.

LENGTHS OF PIPE SEWER ONE BARREL OF CEMENT WILL LAY.

Size, inches	4	6	8	9	10	12	15	18	20	24
Length, feet	500	350	200	175	150	100	75	65	60	50

The following gives approximately the lowest practicable cost of excavating trenches in compact loam or material excavated with equal ease. The prices for shoring and sheathing are to be added where necessary. These are based on continuous work with gangs of the most economical size. House-connections or other short lines would cost more. Profit is not included.

TABLE No. 21.

COST OF EXCAVATING AND BACK-FILLING AND OF SHEATHING TRENCHES.

(Compact Loam; No Ground-water; No Machinery.)

Depth of trench, feet	6	8	10	12	14	16	18	20
4- to 10-inch sewer	.075	.10	.14	.20	.25	.33	.39	.52
15-inch sewer	.09	.125	.175	.24	.315	.39	.49	.65
20 " "	.105	.15	.21	.29	.38	.465	.585	.78
24 " "	.12	.175	.245	.34	.45	.545	.685	.91
30 " "	.14	.20	.28	.39	.50	.62	.78	1.04
Close sheathing. { Lumber, @ $10.	.34	.42	.51	.64	.80	.92	1.06	1.22
Setting	.08	.09	.11	.13	.20	.22	.24	.26
If used 2½ times *	.22	.26	.32	.38	.50	.58	.66	.75
If sheathing-planks 4 feet apart	.10	.12	.15	.18	.24	.27	.30	.33

* ½ used the second time, ⅓ the third time, ¼ the fourth time. With care good sheathing may be used an average of five times.

QUICKSAND may cost from two to ten times the above. No estimate can be given for it. ROCK EXCAVATION in sewer-trenches usually costs from 75 cents to $2 per cubic yard. The greater the amount of rock per running foot to be excavated the more cheaply it can be done.

The approximate cost of laying sewer-pipe, including all but the excavation, is given in the following table:

TABLE No. 22.

COST OF LAYING SEWER-PIPE, CENTS PER LINEAL FOOT.

	2-foot Lengths.			3-foot Lengths.										
Size, inches	4	5	6	8	9	10	12	15	18	20	24	27	30	33	36
Unloading, hauling and distributing.*	0.15	0.18	0.21	0.35	0.38	0.41	0.63	0.90	1.20	1.50	2.10	3.00	4.00	4.46	5.45
Laying, cost of jute, calking	1.23	1.38	1.55	1.02	1.78	1.97	2.31	2.58	3.00	3.70	4.50	5.06	5.48	6.00	6.80
Cement, mixed	0.44	0.50	0.55	0.58	0.65	0.75	0.93	1.11	1.12	1.35	2.00	2.35	2.90	3.86	4.17
Total	1.82	2.06	2.31	2.55	2.81	3.13	3.87	4.59	5.32	6.75	8.00	10.41	12.38	14.32	16.62

* Teams hauling 4500 pounds per load; average haul one mile; $3.50 per day.

The approximate cost of building circular brick sewers is given in the following table. This does not include excavation or back-filling.

TABLE No. 23.

COST OF CIRCULAR BRICK SEWERS PER LINEAL FOOT.

Interior Diameter, Feet.	One Ring.		Two Rings.			Three Rings.	
	2	3	3	4	5	5	6
Brick, @ $8 per M	.40	.58	1.31	1.71	2.10	3.36	3.95
Mixed ⎰ Cement, @ 80 cts. per bbl..	.075	.11	.23	.29	.36	.58	.70
1 : 2 ⎱ Sand, @ 50 cts. per yd	.015	.02	.04	.05	.07	.11	.13
Masons, @ $2.50	.10	.14	.30	.37	.45	.72	.92
Helpers, @ $1.50	.15	.22	.45	.58	.70	1.08	1.43
Total	.74	1.07	2.33	3.00	3.68	5.85	7.13

The approximate cost of manholes, 3 feet by 4 feet 6 inches on the bottom, is given in the following table. A 4-foot circular manhole will cost about 4 per cent more. This table does not include the cost of the iron-work. The brickwork is taken as 8 inches thick down to a depth of 12 feet, and below this as 12 inches thick.

TABLE No. 24.

COST OF MANHOLES, 3 FT. × 4 FT. 6 IN.

(Brick 2 in. × 4 in. × 8 in.; ¼-inch joints.)

Depth, top of Brick-work to sewer-invert.	8 ft.	10 ft.	12 ft.	14 ft.	16 ft.	18 ft.	20 ft.
Brick, @ $8 per M	10.36	12.70	15.04	19.00	23.25	26.75	30.15
Mixed ⎰ Cement, @ 80 cts. per bbl..	1.48	1.82	2.15	2.72	3.32	3.82	4.31
1 : 2 ⎱ Sand, @ 50 cts. per yd	.26	.33	.38	.49	.58	.68	.75
Masons, @ $2.50	2.00	2.25	2.90	3.50	4.45	5.15	5.75
Helpers, @ $1.50	2.40	2.70	3.48	4.20	5.35	5.50	6.00
Total	16.50	19.80	23.95	29.91	36.95	41.90	46.96

Foundations of concrete 6 inches thick, with benches for 8-inch pipe $3.25
Cast-iron tops and covers, 450 to 800 lbs., @ 1¼ to 2 cts......... $5.60–$16.00
Steps, wrought-iron, each.................................... 20 cts.

In the above tables labor is taken at $1.50 per day, teams at $3.50. The cost given does not include superintendence, use of tools, profit, or any of the general expenses of management, but is thought to be liberal, and sufficient to include these under good management.

Natural CEMENT can be had in the Eastern States at from 65 cents up, Portland from $1.80 up, per barrel in car-load lots.

The cost of SAND will vary with the locality from 25 cents to $2 per cubic yard.

ART. 58. METHODS OF ASSESSMENT.

While not an engineering feature of sewer construction, the methods employed in paying for a system may be briefly considered to advantage. In many cases the city pays for the construction and later reimburses itself by special assessments on benefited property or by annual rental; in some the entire cost is borne by the city at large; in others part is borne by the city, part by the property-owners. The city's payment may be made from the ordinary funds or by issuing sewerage bonds. In Philadelphia, Pa., the assessment bills are assigned to the contractor, with right of lien for collection.

Probably no better general statement of sewer-assessment methods in this country could be given than by quoting from an article on the "Theory and Practice of Special Assessments," by J. L. Van Ornum in volume XXXVIII of the Transactions of the American Society of Civil Engineers (September, 1897):

"In a majority of cases the city pays for main sewers, either wholly or all above the usual assessment for a branch sewer. A large number also assess this expense by the area method upon the property affected, either entirely or all exceeding the usual charge for a lateral, as before. Less commonly a percentage is assessed and the city pays the balance, or the cost is divided between an area and a frontage charge, or other plans are followed in its distribution. Of the methods pursued in providing for the collecting system, consisting of the laterals or branch sewers, a plurality prefer to charge the cost upon abutting property according to the frontage rule; though nearly an equal number have an arbitrary rate per foot front, varying from 30 cents to $2, the city to pay the balance; and a considerable number assess the cost either upon the drainage-district or upon a zone of a certain width on each side of the sewer, in the ratio that the area of the lot or land in question bears to the total assessed area, streets being excluded. Of the remaining methods some divide the expense between the city and private property by various processes, others charge it upon property by a combination of the frontage and area rules, and sometimes the city bears the whole cost.

"The frequently occurring plan of assessing upon contiguous property the equivalent expense of a sewer of small size, where a large sewer is placed, is commendable. This method would obviously have no advantage where the total cost of both mains and branches are together distributed *pro rata* upon all the property benefited, nor any application where the city pays entirely for its sewer system; but where adjacent property is charged with a certain part or all the expense of the sewer, inequality would result if the method just indicated, or an equivalent specified fixed charge not depending on the size of pipe, is not applied. Necessarily the larger sewers are laid on the lower ground, where, except manufactories and similar industries, the less valuable and productive property occurs. Here, also, are more generally found tenements and the habitations of laboring men who are less able to meet the burden, while the commercial districts, and especially the dwellings of the more prosperous, are in the higher portions of the city, where the sewers are naturally of smaller size. The latter classes of citizens make the greater use of sewers, and it would manifestly be unjust to fail not only to lay upon them an equal burden, but to charge them even a smaller amount than the average. The cost of appurtenances, like manholes, lamp-holes, catch-basins, and flushing-tanks, is sometimes met by the city and sometimes included in the cost of the sewer and so distributed. The disposition of these expenses depends upon the provisions of law. House-connections with the sewer are made at the expense of the property. In addition a few cities impose a special charge for the privilege of connection for the purpose of increasing the sewerage fund of the city, but this is to be deprecated as tending to discourage the general use of the sewers, which has become a sanitary necessity in cities.

" The question of the distribution of the cost of a sewer system is also a complicated one, whether considered in the light of practice or principle. All the city has an interest, both general and sanitary, in its sewers, and the property-owners have a direct interest as abutters as well as a particular, but more general, one in the larger mains of their district sewers. As far as the trunk sewers are concerned, their construction is of more general import to the city as a whole than to any individual users, and their cost might well be paid by general taxation. Whether or not the city's share in building sewers should always be devoted to these mains, because they have the least direct connection with property, may be uncertain, as custom or local usage may dictate the assumption of the cost of work on street intersections or in front of city lots, parks, and other property, or other expenses, besides the occasional defaults that come upon the city, all of which would probably equal the proportion suggested. All the reasons already given for considering it equitable for the city to share in the expense of its water-works system apply equally to its sewer system; where there are no storm-water sewers (a separate system) for which the city usually pays, it is but just that the city should aid in the construction of the more usual combined system, which has to receive the storm-water from the streets. It would be unfair to expect lots or lands so distant that they may not be able for years to secure connection with the system as it develops to contribute much toward paying for trunk sewers which will at best be of only indirect special advantage to them; and it is believed that the city assuming a share, to the extent of 20 or 30 per cent of the cost of its sewer system, would furnish but a fair equivalent for its benefit, and make less burdensome the individual assessments which so frequently cause objection and retard the construction of these necessary improvements.

"Of the methods followed perhaps the most adequate plan of dealing with the portion of the expense of sewers that is to be assessed is that common one of considering together all the sewers of each sewer district and distributing the cost over the district in proportion to the advantage received. In many cities this allotment is attempted by the frontage rule, but deep lots generally have a larger share in the use of sewers than have shallow ones of the same frontage. The amount of storm-water to be removed from lots is far from having a definite relation to frontage, and other irregularities result. Other cities apportion this assessment by the area rule, but of equal areas that which has the greater frontage enjoys conditions favoring a larger number of buildings or other improvements which imply a greater interest in the sewer system, and therefore should furnish a correspondingly larger contribution; and as systems are often built a portion at a time, lands remote from the constructed portions should not be required to pay equally with lots that are enabled to make use of them at once.

"In consequence of the inequitable features inhering in both systems, in numerous instances it has become an improved method to combine the two processes and assess 40 per cent, more or less, by frontage and the balance by the area rule, or to apply some equivalent procedure that will effect a similar combination of methods. This system of apportionment is growing in favor. It corrects the more serious errors of either method used alone. It is not complex in application, and in principle it is as definite and as easily understood by the people affected as either single process. Probably no more adequate plan for sewer assessments has been extensively used than, after the city has contributed its due portion, assessing by frontage an amount equal to the cost of a smaller size of pipe upon abutting property, as previously mentioned, or an equivalent amount, and distributing the remainder in proportion to area.

" In some Massachusetts cities the plan has recently been applied of partly paying the cost of the sewer system and its maintenance by a sewer rental corresponding in its principle to the water rates of water-works systems. The private contribution to sewerage construction should correspond very closely to the use made of them; and to effect this Brockton and other Massachusetts towns have adopted the plan of such an annual charge depending upon the amount of water used, claiming that the quantity of sewage to be disposed of can be approximately estimated by reference to the water rate. If this plan does not tend to discourage the use of sewers, if it does not too much complicate the system of assessment and proves otherwise practicable, it may furnish a valuable addition to the methods of apportionment. Its practical operation will be watched with interest by those making a study of special assessments."

Since the assessment is strictly a legal function, the State laws and city charter will to some extent control the methods in each particular case. For instance, in New Jersey assessment by the front foot has been decided illegal. It is probable that in all States it is legal to assess in proportion to benefits received, and this too would seem to be demanded by fairness. These benefits consist of: (1) removal of house-sewage from the buildings; (2) removal of rain-water from premises and streets (where combined or storm sewers are built); (3) draining the land when wet; (4) increasing the value of property; (5) a general public benefit to the entire city, whether it is all sewered or not, consequent on the improved healthfulness of certain sections, on increased valuation and therefore reduced tax rates, and on the recommendation to prospective residents that it "has sewers." The 1st, 3d, and 4th are individual benefits, the 2d a combination of individual and general, the 5th a general one. It would appear, therefore, that a perfectly just apportionment of the cost would assess part, but only a part, of this upon property directly benefited, and this at fixed rates in proportion to front foot and area combined, with an additional charge for each connection made or an annual rental in lieu thereof. This charge may be collected either as an assessment, to be paid at once or to bear interest and be apportioned into several annual payments; or in the form of rentals, at fixed rates for the different classes of sewage-contributors or else proportioned to some function of the water-consumption. The method employed at Malden, Mass., of giving each property-holder the option of either of these methods has some advantages. It is in most cases desirable to make each assessment or rental operative as soon as it becomes possible to connect the property with the sewer and make use thereof, as tending to hasten the general use of the system. Whatever the method it should be simple of application and readily understood, should not be burdensome to the poorer property-holders, and should encourage early and general use of the sewers.

For descriptions of various methods employed see the paper quoted from above, Report of the Engineer to the Brockton Sewerage Commission, and Journal of the Association of Engineering Societies for January and March, 1897.

PART II.

CONSTRUCTION.

CHAPTER X.

PREPARING FOR CONSTRUCTION.

ART. 59. CONTRACT WORK OR DAY LABOR.

THERE are two general plans by which a city or town may construct a sewerage system, viz., by contract or by day labor. In a majority of instances, probably a very large one, the contract method is adopted, but in quite a number the work is done under the general charge of the city engineer or a special agent or committee who purchases material, employs labor, and looks after the work generally. If the work can be kept entirely free from politics and conducted without "fear or favor" by a good manager experienced in this line of work the latter method will probably be the more economical for the city and give the more satisfactory results. Unfortunately these conditions exist in few cities or towns, and the contract method is usually the cheaper one, and frequently gives better results than construction by home labor under foremen too often unskilled in sewerage-work. There may be cases where, even with and in spite of the existence of the above objections, construction directly by the city is preferable. For instance, the work may be of an uncertain nature, its details difficult to foresee and set forth in a contract; or it may be unusually hazardous, causing contractors to add 100 or even 200 per cent to the estimated cost to balance the risk, which risk the city might think it better to assume itself. In some instances villages have undertaken sewerage-work as a means of giving employment in unusually hard times to citizens, who would thus be enabled to pay part of their wages back to the treasury in taxes, and also relieve the poorhouse of a large number of possible inmates.

Since, however, sewerage-work is generally done by contract, the succeeding chapters will be made applicable particularly to construction by this method. But the matter is equally applicable to work done by the city or its immediate agent, which agent should conduct the work as the contractor would have been compelled to conduct it.

It may be sometimes advisable for the city to purchase the materials and contract for the labor of construction. In this way the quality of the materials is under the immediate control of the city. In the matter of cost there is usually very little difference one way or the other, unless it be that the city is charged a little more, owing to a "commission" which must be paid to certain officials who control the awarding of the contracts for material. It is an excellent plan for the city to furnish cement, sand, and pipe, and see that there is no unnecessary waste of these. There is then no temptation for the contractor to use defective material or too little cement.

Systems have been built by letting the contract for excavation to one party, and that for pipe-laying and brick-work to others, the material being purchased by the city. This is almost sure to work unfavorably to the city and give rise to the greatest confusion, of responsibilities if not of work.

Art. 60. Obtaining Bids.

If work is to be let by contract the probabilities are that the greater the number of bidders the lower will be the sum for which the work can be constructed, a partial cause of this being the lessened liability of collusion between the bidders. To reach contractors two methods are open: to send notice to individual contractors, or to advertise in such a way and place as will attract the attention of a large number. Each method has its advantages, and perhaps a combination of the two might give the best results. But it is probable that one or two advertisements judiciously placed will reach all who would be reached by the first method, and many others besides. For a small village the best advertising medium is the contractors' journal having the widest circulation in that part of the country, which is also true for a city if the work amounts to more than a very few thousand dollars. For small contracts in cities having several capable contractors among its citizens the local paper will perhaps give sufficient publicity to the desire for bids, but village papers are generally useless for this purpose. For contracts of $5000 or more an advertisement in one or two prominent engineering and contracting papers will usually pay for itself many times over.

The customary method of bidding is to have each contractor submit a sealed proposition, all of these being opened at a time fixed beforehand. It would be unfair, if not illegal, to open any sealed bid before this time or to receive another bid after any had been opened. To satisfy both the public and the contractors of the honesty and fairness of the awarding of the contract it is customary to open the bids and read them aloud at a meeting open to the public, or at least to all bidders and to newspaper representatives. The opportunities for dishonesty are so great if the bids are opened in secret or one received after others have been read that regard for their own reputations usually influences the officials making the award to adopt the above methods.

That there may be no informal or incomplete bids it is desirable that all bids be made on forms furnished by the city and accompanied by copies of the specifications and contract similarly furnished. It would be possible for a bidder whose bid had been accepted to refuse to contract for the work unless he were bound in some way. For this reason it is well to require that each bid be accompanied by a certified check, to be returned to the bidder unless he refuse to sign the contract based upon his bid, if so requested.

The laws of different States and cities differ as to the latitude given in awarding contracts. In some the contract must be awarded to the " lowest bidder," in others to the " lowest responsible bidder," while in still others there is no restriction. Justice to the taxpayers and fairness to the bidders will usually dictate awarding to the lowest bidder, unless there be reason to think that he will be unable, through inexperience, to do creditable work, or that, his bid being lower than the work can probably be done for, he will later abandon the work, and the consequent delay and legal complications, even though his bond insure the ultimate completion of the contract, will be detrimental to the city's interests. If it becomes evident during construction that the contractor cannot but lose money there is usually a tendency to favor him in minor matters, to grant him extensions of time and aid him in other ways which detract more or less from the excellence of the work. In order to avoid, on the other hand, the necessity of awarding the contract to a too high bidder when there are no reasonably low ones the city should " reserve the right to reject any or all bids."

Art. 61. Engineering Work Preliminary to Construction.

As soon as possible after the signing of the contract the contractor should submit samples of the material he wishes to use, and these should be carefully examined by the engineer, and if accepted should be retained and marked for future identification and compared from time to time with the material actually furnished.

The contractor should be notified some days in advance of the point or points at which he is to begin work. Reasonable deference should be made to his wishes in this matter, since it is his privilege and duty to so organize the work as to secure the greatest efficiency at the least cost to himself. If, for instance, part of the work lies through wet ground and sub-drains are to be used it is ordinarily to his interest, and indirectly to the city's also, that the work begin at an outlet to which all ground-water will drain, or at a point at which a pump, once set up, can drain the work for long distances without moving its location, as at the junction of two mains. It is also usually desired by the contractor that, if two or three gangs are to work at as many places, they may be within a few blocks of each other for convenience of oversight. It will ordinarily be to the interests of both contractor and city to work in as dry ground as possible, and hence to leave until summer droughts construction through low, soggy land. Construction across or near streams should not be carried on when there is a possibility of floods or freshets, if it can be avoided. Both trench- and masonry-work should be avoided in winter weather if possible, for it is then costly to the contractor, and it is impossible to be sure that the mortar is uninjured, or to restore the streets to good condition with frozen earth.

Ordinarily the contractor will desire to place upon the street, along the line of the work, pipe, brick, sand, lumber, etc. This cannot be denied him, but he should be compelled to place and pile this material so as to interfere with travel as little as possible, and along only those stretches of street in which construction is to be begun within a week, or ten days at the outside. This material should be inspected as it is delivered, and that condemned removed at once.

Just before the work begins it is well to run levels carefully over all bench-marks to see that they have not been disturbed and to check previous levelling; also to establish new ones if necessary. It is desirable to so place these that one of them can be seen from the instrument when set up for giving grades to any part of the work. They should be accurate within at least .003 of a foot.

Art. 62. Other Preliminaries.

Final arrangements should now be made for the oversight of the work, the proper instruments obtained, engineering and inspecting assistants engaged, an office or other headquarters arranged for, notebooks and blanks obtained for making and preserving records, final arrangements made as to right of way across private property and along county roads or others not controlled by the city or village. Arrangements should be made also for locating the branches for house-connections at the points desired by the property-owners. For this purpose it is well to publish in the paper or otherwise make known to the citizens that each is desired to drive, at his fence-line or curb, a stake indicating the point at which he wishes his house-connection to enter his property, and that in case no such stake is driven the engineer or inspector will use his judgment in locating such branches. Another method is that of requesting that sketches of the property showing such point be handed in on blanks to be furnished by the engineer; but the inability or hesitancy of many citizens to make the simplest drawing is an objection to this plan.

Counsel for the city should pass upon the sufficiency and correctness of the contracts signed, of the bonds given and of their signers, and all other legal matters in connection therewith, before the contractor is permitted to begin work.

CHAPTER XI.

LAYING OUT THE WORK.

ART. 63. LINING OUT TRENCHES.

SINCE the trench is seldom more than 6 inches wider on each side at the bottom than the sewer to be placed in it, it is necessary that the trench itself be carefully aligned, and this cannot be entrusted to the contractor except for short distances. For giving him the line the safest plan is to drive stakes or spikes along the centre of the proposed trench at intervals of about 50 feet. To assist in finding these, for checking, and to locate the centre of the sewer during construction, the distance should be taken from each of these to the curbing opposite, if there is any, or to a reference-stake, and a note made of this. The centre spikes or stakes should be some uniform distance apart to facilitate finding them. They should be set by a transit placed over the centre of a manhole on the line. The line should of course be straight between manholes, except in the case of large sewers, which may be curved, in which case the centre stakes should be set 10 or 15 feet apart. The location of manholes, flush-tanks, etc., should be fixed by two or more reference-stakes and pointed out to the contractor before he begins excavating, that he may make allowance for them in sheathing.

Some engineers in giving line rely entirely upon reference-stakes placed a uniform distance from the centre of the trench, because the centre stakes will be removed in excavating. An experience with the unreliability of contractors' tapes and foremen's intelligence seems to argue in favor of the centre stakes, however.

It is well to do a large part of this lining out before construction gets well under way, since it is probable that the engineer corps will be kept busy with other work later. Each line should be located only after due consideration of the points referred to in Art. 34.

ART. 64. GIVING GRADE.

Several methods of giving grade are employed by different engineers, the principal being: By means of a cord stretched over the centre of the trench and parallel to the sewer grade; by stakes driven to grade along the centre line in the bottom of the trench; and by stakes driven at the ground-surface near the edge of the trench, their tops a uniform or stated distance above the sewer grade. (The grade used in both designing and construction is that of the inside bottom of the invert, which for convenience will be called the invert.) Each of these is used for both pipe and brick sewers, but only the first method is at all adapted to accurate laying of pipe sewers. For brick sewers either method may be used, but the first is most convenient in that the invert-templet can be set at any point along the trench, and that the bottom of the trench can be carried to the exact grade at every point, in advance of setting the templet, by measuring down from the cord. If stakes are driven in the bottom the templets can be accurately set only at points close to these, and the stakes can be driven only when the bottom is within a foot or less of grade, which necessitates the presence of the engineer upon the work almost constantly. If the stakes are driven along the edge of the trench they can be set even before the excavation has been begun, enough for several days in advance being set at one time; but it is almost impossible to avoid errors in measuring down from these, since they are not directly above the sewer, and the stakes are apt to become loose or fall out with the cracking and caving of the edge of the trench.

The cord used in the first method may be fastened to a strip of wood nailed in a vertical position to a plank which stands upon edge with one end resting upon the ground on each side of the trench. This plank should extend at least 18 inches or 2 feet beyond the trench on each side and be firmly bedded into solid ground so that it cannot possibly settle, and should be held upright on edge by a stake driven on each side at each end, or by stones and earth solidly banked around the ends. These grade-planks should be not more than $33\frac{1}{3}$ feet

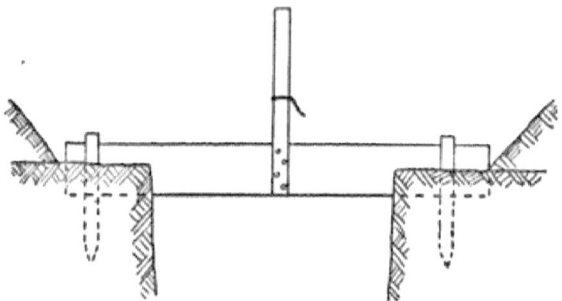

FIG. 5.—METHOD OF SETTING GRADE-PLANK.

apart—25 feet would perhaps be better, since the cord will sag too much if the distance between supports be greater. On the top edge of these planks the centre of the trench is marked, and strips of wood about 1 inch × 2 inches × 24 inches are nailed so that one edge of each is in this centre line and truly vertical, as determined by a plumb-bob. On this edge is placed a mark exactly a whole number of feet above the sewer-invert immediately beneath, and a slight notch is cut to receive the cord. All notches in a given length of sewer are placed the same distance above the sewer-invert, and the cord stretched from one to the other is therefore parallel to the grade, which can be found at any point by measuring the given fixed distance down from the cord. The cord is also vertically above the centre line of the sewer. If the trench changes in depth or for some reason it is desirable to change the distance from the cord to the sewer-invert, a step up or down must be made at some grade-plank by cutting two notches one or two feet apart in elevation. The cord should be strong linen fish-line or similar material whose light weight will prevent unnecessary sagging, and should be stretched tightly between the grade-planks.

Another method of supporting the string is to drive at equal intervals stout stakes, at least 2 inches × 4 inches × 5

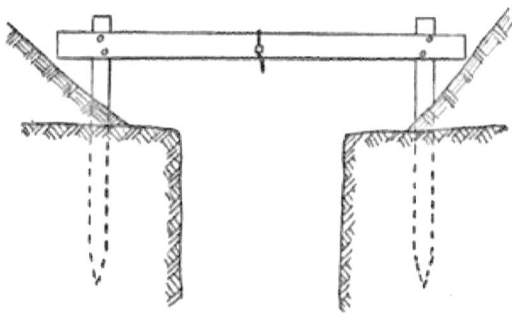

FIG. 6.—METHOD OF SETTING GRADE-PLANK.

feet in dimension, on each side of and about 3 feet back from the trench, and in pairs directly opposite each other. On each of these is found a point a certain whole number of feet above the sewer-invert and usually 2 or 3 feet above the ground, and a straight-edged board or plank is nailed, one end to each stake, with its upper edge exactly at these points. On this edge is marked the centre of the trench and a slight notch cut there to receive the cord. One end of the cord is fastened to a nail driven into the first board below the notch, and rests in this notch and those of succeeding boards, and is fastened at its other end to another nail placed as is the first. Or a large spool is cut as shown in Fig. 7 and caught behind the board, the cord being fastened in it and being readily tightened whenever it becomes loose. This method of elevated grade-boards is particularly applicable to large pipe sewers or small brick ones, since, the cord being higher above the ground, it interferes less with the lowering of materials into the trench. In some cases it is not wise to adopt it on account of the liability of the banks to be caved in by the driving of the posts.

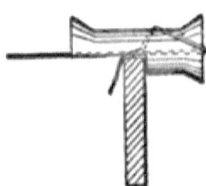

FIG. 7.—METHOD OF HOLDING GRADE-CORD.

In laying pipe sewers from the cord a grade-rod is used with a mark or notch on its edge so placed that when it is level with the string the foot of the rod is level with the sewer-invert or with the outside top. The former is preferable, since the invert is the part of the pipe which it is most important to have at correct grade, and, as the pipes often vary slightly in diameter, this result may not be obtained if they are graded by their tops. If the inverts are to be set it will be necessary to have an offset-piece at the foot of the grade-rod which can rest inside the pipe upon the invert. For this purpose an ordinary cast-iron 6- or 8-inch bracket, obtainable at any hardware-store, will answer; or wrought iron may be used if about $\frac{1}{4}$ inch thick and stiffened by being bent back upon the rod. The mark on the grade-rod should be checked each day.

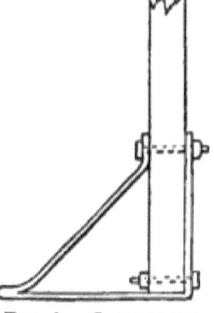

FIG. 8.—GRADE-ROD.

For brick sewers each templet is set by a rod, and for both these and pipe sewers another rod is used by the foreman for getting the excavation to the proper depth.

The grade-plank or stakes above described can be set out even before excavation is begun, but except in shallow trenches it is better to wait until the trench is at least 6 feet deep, that they may interfere with the excavating as little as possible. It is often well, however, to drive the stakes, where these are used, before excavating to prevent cracking the bank, the board not being nailed to them until afterward. The grade-plank and stakes should be tested for grade and line at least once a day, and the inspector should keep close watch of them to see that they are not disturbed and also that the cord is kept taut.

When excavating-machinery is used the grade-planks cannot ordinarily be placed above the surface, but they can be sunk into the ground entirely below the surface, or the bracing of the trench can be utilized by nailing the vertical strips to it. This latter method is also preferable for pipe sewers when the trench is very deep, since, if the cord is at the surface, the grade-rod is too long for convenience and accuracy, and the inspector is too far from the work of sewer construction to watch properly both it and the grade-rod.

In trenches through running- or quicksand unless the utmost care is taken the banks will settle several inches or even feet, carrying the grade-plank with them of course. Under such circumstances a level should be kept constantly on the ground and the grades checked every few minutes during pipe-laying or at the time of setting templets. Moreover, the bench-marks themselves, when on curbing, fire-hydrants, or elsewhere near the roadway, may settle and should be checked daily.

For properly fixing the grade of a manhole-head a stake may be driven near by indicating the street grade, and it would also be well to test the head by the level as it is being set. Similar stakes should be set for storm-water inlets.

Inlet- and house-connections should be laid as truly to line and grade, and in the same way, as the sewer itself.

Where grade-stakes in the bottom are used these are set to the exact grade of the invert or one foot above it. For pipe sewers the bottom of the trench is then given a uniform grade from one stake to another by using a straight-edge, stretching a cord, or too often by eye only, and the pipe is laid on this bottom and lined in by eye. Great accuracy can hardly be expected by this method. For brick sewers, however, it compares favorably with the cord for accuracy (but not for convenience), the stakes being driven in the centre line of the sewer and at such distances apart that each templet can be set close to a stake.

Stakes driven upon the bank are not recommended for any purpose, it being almost impossible to obtain accurate results by their use. Their only advantage is that setting them gives less trouble to the engineer than either of the other methods.

Alleged sewers have been laid with a carpenter's level 2 feet long, to the under side of one end of which was fixed a piece of wood or iron or a screw protruding an amount equal to the desired rise in that distance. It would be an exceptional case in which a line of sewer so laid did not vary more than one inch in each 100 feet from the desired grade.

While giving grades the measurement of the sewer-lines should be carefully made and noted and compared with the original measurements, and if any appreciable difference is found the sewer grades upon the plans must be readjusted to correspond. Careful notes should be kept of all instrumental work connected with giving line and grade. It will be convenient to have in each level-notebook a list of all benchmarks in the sewer-district in which it is to be used.

Both inspector and engineer should watch for the first indication of the existence in the trench of an obstruction to the sewer, that preparation may be made for a change in line or grade if necessary to pass the obstruction. Such change, if in line, may necessitate inserting one or two additional manholes or a lamp-hole; if in grade it may be made by a decrease in grade in one stretch and an increase in the next, or by siphoning under or over the obstruction (see Art. 49). In some cases the change can be made in the obstruction and not in the sewer.

It is the inspector's duty to see that house- and inlet-connection branches are inserted at the proper points, and their exact locations noted, which locations the engineer must make note of and reference to some fixed point, usually the centre of the nearest manhole, to make possible the ready finding of the branches in the future. This is very important and should be faithfully attended to.

CHAPTER XII.

OVERSIGHT AND MEASUREMENT OF WORK.

ART. 65. INSPECTION OF WORK.

THE specifications are practically the instructions to the contractor as to the way in which the sewerage system is to be built, the lines, grades, and dimensions being given by the engineer, chief or assistant. If the contractor were left unwatched to carry out these instructions it would be impossible to know whether he had done so or not, since only the inside of the sewer can be examined, and this only with difficulty. And if it were found, after the completion of the work, that it had been improperly built or of poor material, even though the contractor could be compelled to replace it with satisfactory work, the delay and inconvenience of this might better be avoided by proper oversight during construction. It is advisable, therefore, that a competent inspector be constantly on hand when any construction is progressing. This is not necessary during excavation, but even this should be looked after at least once a day, that any unforeseen underground condition which may modify the plans may be noted, and in general to ascertain that the contractor is obtaining the proper width of trench, is not interfering unnecessarily with private drains, water- or gas-pipes, and is in general following the directions for trenching, blasting, etc.

For this oversight it will usually be necessary to have an inspector for each set of pipe-layers and of masons. But if only one or two trenches are being worked at a time the instrument-man may also be inspector. The omissions and poor work which may be accepted from the contractor if such inspection is not constantly made may be seen from a statement of the inspector's duty.

The inspector should be on hand before work is begun at morning and noon to see that no mortar left from the previous day is worked over, that new mortar is properly proportioned and mixed, and to examine grade-lines or -stakes. In the case of pipe sewers he should examine the inside of the sewer near the end and see that any stones, dirt, or other matter which may be there be removed before the laying begins. He should also examine the one or two cemented joints nearest the end, and if they are not sound the pipe should be removed and relayed. In the case of brick sewers he should examine the toothing at the end of the brick-work and have removed any loose brick and all mortar and dirt that may be lodged there.

He should continually keep an eye upon material and workmanship, examining each pipe before it is lowered into the trench, each load of brick and of sand as they are brought upon the ground, each barrel or bag of cement to see that it bears the engineer's mark or is of the required make and that it is not caked by moisture. He should see that the proper proportions of cement and sand are used for the mortar, and that no mortar partially set is retempered and used.

On brick sewers he should see that each templet used is one approved by the engineer and that it is set to the proper grade and line, that the brick are laid to line and in accordance with the specifications, that slants or other branches are set where needed, and he should keep an accurate account of these, their size and length, and mark the position of each by a stake driven in the bank directly over it for the information of the engineer. He should see that the arch-centre is solid and does not spring under the brick-work, and that it is not drawn too soon.

On pipe sewers he should see that each pipe is laid to grade and line by the use of the grade-rod and a plumb-bob in connection with the grade-cord, that each pipe is pushed "home," each joint properly cemented and the swab or piston in the sewer pulled forward, and that the back-filling is properly placed and tamped around the pipe. He should see that house-branches are placed where directed, that covers are cemented in each one (about this he is sometimes careless, to the great detriment of the sewer), and should drive a stake in the bank directly over each.

He should keep a record of all extra work, or work, such as foundations or sheathing left in the trench, which cannot be measured after the completion of the sewer.

If the ground is wet he should see that no water flows over the brick-work or through the pipe, except as permitted by the engineer. In general he should be thoroughly familiar with the specifications and have a copy constantly on the work, and see to their enforcement, reporting to the engineer any difficulty in obtaining this.

He should not be permitted to be in any way indebted to, or under the influence and power of, the contractor, and should receive orders from the engineer only.

He should be a man with some experience in the character of work he is inspecting, sober, and having the respect of the contractor and workmen.

ART. 66. DUTIES OF THE ENGINEER.

The engineer should keep constantly in touch with the work, visiting each point at least once a day, and giving necessary instructions to the contractor and inspector, as well as giving and testing line and grade. If he has many inspectors on work under his charge they should be required to report at the engineer's office after each day's work the amount done and return a detailed statement of any extra work, asking instructions on any points concerning which they are in doubt. The daily reports may be made in writing upon blanks furnished to the inspectors for this purpose.

The engineer must see that each inspector is performing his duty, and if necessary enforce instructions given by him to the contractor. He must inspect all material to be used, where this is possible, or give the inspector full instructions on this point where it is not. It is well to mark each accepted barrel or bag of cement; to inspect the pipe after it is delivered upon the street, but well in advance of the laying, seeing that all defective pipe is removed; also to inspect and weigh all iron-work before it is used.

The engineer must decide where and how much sheathing shall be left in the trench, making a note at the time of its exact location and length, must decide as to the classification of the material being excavated, and must measure promptly all material classed as rock. He should each day take measurements necessary to locate the house-branches as indicated by the inspectors' stakes. It is well to measure each stretch of sewer, each manhole and other appurtenance as soon as completed.

The engineer should see that the contractor respects the rights of property-owners and keeps the streets and sidewalks open where possible, that the laborers are efficient and, where necessary, skilled in the work to which they are assigned, and that they create no disturbance along the streets in any way for which the contractor is responsible. He should compel the contractor to work with sufficient force and in such a manner as will lead to the completion of the work in the specified time, to place such shoring and sheathing as may be necessary to prevent any accidents to property or lives or to the sewer, to provide pumps in sufficient number and of ample size to handle all ground-water, and to use excavating-machinery where necessary. In general he should see that the work is carried on by proper methods, with proper materials, with a force and a plant satisfactory in both character and extent, and that the inspector enforces his directions as to details.

Art. 67. Measurements.

The specifications should state in what way the measurements shall be taken for each description of work or material. The measurements so made for the final estimate (which is the name customarily given to the measurements and calculations on which is based the final payment for a piece of work) should be accurately and carefully taken and checked at least once, as should be the calculations based thereon. The engineer should be able to swear upon the witness-stand, as he may be called upon to do, that the final estimate is a truthful and correct statement of the work actually done. Quantities given in this estimate should be stated in the units used in bidding for the work.

Measurement of the sewer laid is usually made from centre to centre of manholes, flush-tanks, etc., not horizontally, but parallel to the sewer (the surface of the street being practically this in most cases), no deduction being made for branch specials or the lengths of manholes. Payment is sometimes made uniformly for all depths of sewer, sometimes varying with varying depths. The latter seems the fairer way, particularly where some lines contracted for may be omitted or new ones added. Usually no changes in price are made for less differences in depth than two feet, the measurement being made from the surface to the under side of sewer or of foundation-platform. These depths are ascertained from the profile, on which are plotted the surface grade and the sewer. Since for the original profile elevations were in general taken at 100-foot intervals only, and as a check on these, the elevation of the surface should be taken and noted at each grade-plank when grade is being given for sewer construction.

The depth of each manhole, lamp-hole, and other appurtenance should be obtainable from the profile, but as a check each should be measured with a levelling-rod or tape. Each manhole, lamp-hole, flush-tank, and inlet should be designated by a number, by which it is referred to. It is almost impossible otherwise to correctly count and keep track of these, especially the manholes, so many of which are each common to two lines of sewer.

Inlet-connections may be measured from their upper end to the shoulder of the branch or slant. Whatever the limits to be taken they should be carefully set forth in the specifications.

Rock should be measured before excavation in most instances, although its original surface can often be judged afterward by that showing along the sides of the trench. If the rock-surface is fairly even and uniform readings may be taken at intervals of 10 feet; but if it be uneven and jagged these should be, not at regular intervals, but wherever necessary to give accurate results. All measurements, whether of earth, rock, sewer, or manhole, should be taken to tenths of a foot. It is customary to allow the contractor a certain cross-section of trench, and pay him nothing for excess excavation nor deduct for a less area of section. But the trench at the bottom should be kept the full width called for.

A final-estimate book should be kept, in which is entered an exact statement of each piece of work as it is completed, but not before then. The measurements should be classified under the items for which bids were received, and the location of each given; thus:

8-INCH SEWER, 8 TO 10 FEET DEEP.

Location.	Length.
From manhole No. 7 to manhole No. 8	327.3 feet
Between manhole No. 8 and manhole No. 9	39.0 "

8-INCH × 4-INCH Y BRANCHES.

Location.	Number.
Between manhole No. 7 and manhole No. 8	13
" " " 8 " " " 9	11

MANHOLES.

No. of Manhole.	Location.	Depth.
7	Main Street, between Clinton and Madison	9.2
8	Corner Clinton and Main streets	9.2

A pocket field-book should be constantly at hand, in which are entered all measurements taken, the points of the beginning and ending of " sheathing left in trench," of sub-drains, of foundations, the location of all Y's, the details and quantities of " extras," the location of underground structures for future reference, and the date of beginning and ending of construction on each stretch of sewer. These notes should be copied every evening into an office-book, since a loss of these data would be serious and irreparable. The general appearance of such notes is shown on page 259.

It is well also to have a pocket copy of the profile of each street, showing the sewer as designed, with size, grade, elevation, location of manholes, flush-tanks, and other appurtenances. This method of taking these data to the field for use seems to be more complete and convenient than copying them down into a notebook.

According to most contracts the contractor must be paid monthly, and for this purpose monthly estimates must be made by the engineer. He should estimate each month the total amount of each item completed to date, from which is deducted the total estimated the month previous, the difference being the amount performed during the month. This method prevents the carrying ahead or accumulation of any errors which may be made in any one monthly estimate, which errors are liable to occur owing to the fact that such estimate

OVERSIGHT AND MEASUREMENT OF WORK.   259

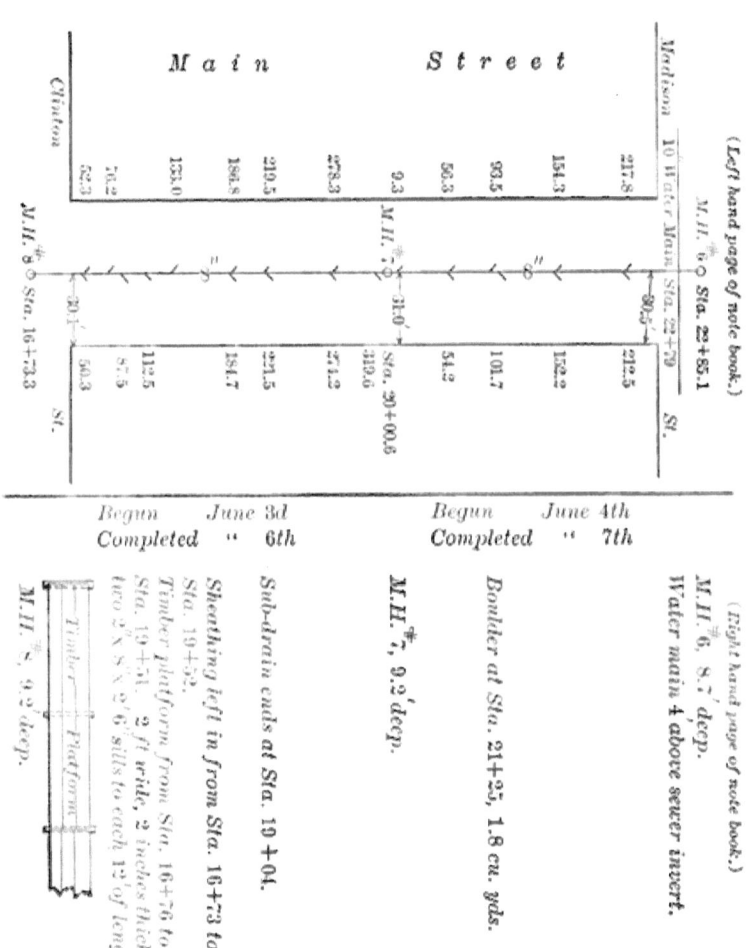

must often be made hastily and simultaneously with the oversight of construction-work. Uncompleted work must be estimated according to the judgment of the engineer as equivalent to so much completed work of the same class.

For the final estimate all measurements as given in the final-estimate book should be checked with the field-book and in every other way possible, and every precaution taken to secure absolute absence of error in measurements or calculations. As a check upon the estimate it would be well to obtain from the contractor the bill of pipe, brick, and iron used by him upon the work, allowance of course being made for material condemned or unused.

ART. 68. FINAL INSPECTION.

The final inspection of the work before its acceptance from the contractor should be thorough, and made by the engineer in person or by an experienced, trustworthy assistant. He should enter every flush-tank, manhole, inlet, or other appurtenance sufficiently large for this, taking its dimensions, noticing whether the head or grating is at the proper level and substantially set, the brick-work smooth, the form regular, the steps properly set at the prescribed intervals, that no ground-water leaks through the brick-work, that pipes passing through the walls are properly built in with surrounding "bull's-eyes;" that the bottoms of manholes are formed according to instructions, the invert-channel being straight or with a uniform curve, of the proper width, and its grade uniform through the manhole and of the proper elevation, and that the benches have the specified slope; also, if there are sub-drains, the hand-holes should be inspected, and these as well as the manholes should be free from dirt. Lamp-holes should be inspected by lowering a lamp into each and noting whether it is straight and vertical, and by seeing that the heads are set according to specifications and at the proper grade. Flush-tanks should be filled with water and tested for tightness for at least 24 hours, during which time the water-level in them should not lower more than one or two inches. If automatic flushing apparatus is set it should be tested with a stream sufficiently small to fill it in not less than 24 hours. To expedite the test it can be rapidly filled and discharged once to test its proper working, then rapidly filled three quarters way to the discharging-point and the inflowing stream cut down to the rate above mentioned, to see that the siphon does not "trickle," but holds the water until the height is reached calculated to cause a complete siphoning of the water in the tank.

Every foot of sewer and inlet-connection should be inspected. Sewers 24 inches or over in diameter should be entered and each joint inspected, if they are pipe sewers, to see that no jute or cement protrudes into the sewer and that there is no leakage. In case of the former the protruding cement or jute should be removed; and if there is leakage this should be stopped, for which purpose there may be calked into the joint from the inside dry cement immediately followed by jute, cloth, or similar material to hold it in place until set; or wooden wedges, or tea-lead may be used. If these or similar methods fail it may be necessary to uncover the pipe and apply additional cement on the outside, backed and supported by concrete if necessary. Any cracked or broken pipe should be dug up and replaced. The branches should be examined also to see that a water-tight cover is in each one which is not already connected with a house-drain.

If the sewer is of brick the brick-work should be smooth, with struck or pointed joints and without any cracks. To determine whether the form and dimensions are as specified a skeleton templet may be used. If the sewer is circular this may consist of two light rods, each of a length equal to the nominal interior diameter, and connected by a bolt passing loosely through holes at the exact centre of each. One of these rods is to be held stationary across the sewer and the other revolved upon the bolt, when each end of the latter should just touch the sewer through the entire revolution. For an egg-shaped sewer a half templet may be used. Slants or other branches should be examined as stated for pipe branches. Special attention should be paid to junctions of brick sewers to see that the curves are easy and uniform in plan and that the arches are strong and well built. All spalls, bats, plank, and other refuse and dirt should be removed from the sewers. If brick sewers leak the joints may be calked, as suggested for pipe-sewer joints. For such inspections a lantern with a reflector is desirable.

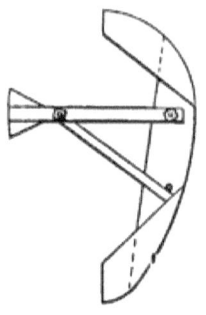

FIG. 9.—INSPECTOR'S TEMPLET FOR EGG-SHAPED SEWER.

Inspection of small pipe sewers can be made from manholes only. As a test for straightness a light held at the opening of the pipe in a manhole or lowered into a lamp-hole should be distinctly visible from the next manhole. Further inspection can be made by the use of mirrors from which light is reflected into the sewer. The simplest plan is to reflect the sunlight from a mirror held by an assistant on the surface to another mirror held by the inspector in the manhole, who so manipulates his mirror as to throw a spot of light onto each length of the sewer in succession, meantime inspecting the same by looking past the mirror into the sewer. This generally requires that he kneel in the manhole upon the side benches, his back to the sewer to be inspected, his head bent down until he can see into the sewer, and the hand which holds the mirror thrust back between his legs. It is advisable that he have pads for his knees if he have many sewers to inspect in this way. Apparatus has been devised for removing some of the inconveniences of this method by so placing an additional mirror that the interior of the sewer is reflected therein, and the inspector is relieved of the necessity of assuming an uncomfortable position. Such an apparatus is described in Engineering News, vol. XXXII, page 249.

The imperfections most commonly found in pipe sewers are: loops or ends of oakum or ridges of cement protruding into the sewer at the joints; dirt, stones, etc., in the sewer; uneven grade, which can be detected by allowing a small amount of water to flow through the sewer, which stream will be wider at the depressed and narrower at the elevated points; ground-water leaking in through the joints; broken pipe; breaks, at joints, in the continuity of the invert-surface.

The last defect can be remedied by relaying the pipe, or by drawing through the sewer an "invert-former" filled with thin neat Portland-cement mortar (slow-setting). This is essentially a box, its bottom of the shape and size of the sewer-invert, through a large opening in which the cement passes to the sewer, to be pressed into shape by the rear end of the box as it passes over it. The box is heavily weighted to force ahead the surplus cement. It can be made of thin

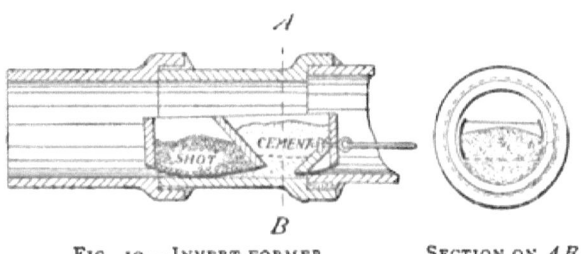

FIG. 10.—INVERT-FORMER.   SECTION ON AB.

sheet iron bent to the proper shape and stiffened by inside partitions cut from plank. Broken pipe and leaking joints can only be repaired by digging down to the sewer (see, however, Art. 85). Dirt, stones, and protruding cement may be removed by drawing a scraper through the sewer by means of a rope, or by pushing it through by a rod formed by jointing together several shorter rods of a size which can be introduced through a manhole—about 5 feet. A stream of water from a fire-hose nozzle under a good head can be used to remove from a stretch of sewer not more than 300 feet long almost anything less in size and weight than a brick. The hose while water is passing through it is so stiff that it can be pushed for a long distance into the sewer. Jute ends or loops are sometimes difficult to remove, but can usually be cut off by a sharp knife-blade fastened to a long rod, or burned off by putting under them (they generally hang from the top of the sewer) a small lamp or candle similarly fastened. Or, if there is water flowing through the pipe, the candle may be fastened to a piece of wood or cork to which a string is attached and floated down to the desired point. The exact distance from the manhole of any defect can be ascertained by counting the number of pipe intervening.

Sub-drains should be inspected by turning into each stretch for a short time all the water it can carry (if they are not already running full) and watching for indications of stoppages. The apparatus for inspecting sewers above referred to may in some instances be used for sub-drains, being lowered into the sub-drain hand-hole. If any drain is entirely stopped this may be remedied by the use of rods, fire-hose, "pills" (see Art. 85), etc.; or it may be necessary to locate the obstruction and dig down to it.

As far as possible assurance should be had by examination that all the conditions of the contract have been carried out, those having reference both to the construction and to the more strictly business relations between city and contractor.

CHAPTER XIII.

PRACTICAL SEWER CONSTRUCTION.

Sewer construction is sometimes undertaken by the city under the immediate supervision of the engineer, who should in such a case be well informed in practical construction methods. He would also be better fitted by such information to design a system and to oversee it if constructed by a contractor. This chapter is intended to give some information on this subject based upon practical experience. It is not pretended that the entire field is covered, but it is thought that the student and those with little experience in sewerage-work, and perhaps others, will find the information given of considerable value.

Art. 69. Organizing the Force.

The number of men which can be worked in one gang economically depends upon the character of soil, depth of excavation, amount of ground-water, manner of construction of the sewer; also upon the personality of the general foreman and contractor. If the soil is "rotten," with little cohesion, very wet, or the trenches very shallow, the gangs should be small; but if the ground is dry and stands up well or the trenches are deep larger gangs can be used. With the increase in the number of gangs comes increased difficulty in keeping them all supplied with materials and tools and working to an advantage. Good foremen are a necessity if there are to be more than two or three gangs, since it may at times be necessary to leave them to carry on their work for days at a stretch with no more than a hasty daily visit from the contractor or general foreman. A foreman who can keep the men faithfully at work without favoritism or making himself generally hated by them, who has sufficient intelligent foresight to arrange their work a day or two ahead, to never be out of sheathing, cement, sand, brick, or other material, who has a practical knowledge and knack for overcoming difficulties, and who can be depended upon to be sober from the time the work starts until it ends—such a man is valuable upon sewerage-work. But if such men cannot be had it will be better to work only two or three gangs, all of which can be kept under the contractor's or engineer's eye.

The city engineer or the contractor, as the case may be, if he does not himself devote his entire time to it, should have a general foreman over the entire work. There should be a foreman over each gang, and if the number in a gang exceeds 30 an additional foreman; also in each a water-boy to carry drinking-water and run errands. If the trenches need sheathing there should be on each, under the direction of the foreman, from one to three men handy and experienced in such work.

It will be necessary to sharpen the picks frequently, even twice a day in flinty hard-pan or gravel, and for this purpose, as well as to repair shovels, wheelbarrows, axes, chains, etc., a blacksmith should be established handy to the work. When not engaged on such repairs he can be making manhole-steps, calking-irons, etc.

There should be a timekeeper, if the force is large, to take the time daily and make it up for each pay-day, who may also serve as clerk, keeping account of all material received and where delivered, ordering new when so instructed, and keeping a daily account of the work done by each gang.

Two pipe-layers may be connected with each gang if the trench can be rapidly excavated; otherwise two or more gangs may have a pair of pipe-layers in common, who lay pipe first in one trench and then in another, as sufficient length of each is excavated. For manholes, flush-tanks, and other masonry appurtenances a mason and two helpers may work together, passing from point to point as needed. For brick sewers two, four, or even six or eight masons may work together, the number in a gang usually being even. The number of masons' helpers depends somewhat upon the depth of the sewer, one or more extra ones being required to lower brick and mortar if the depth is considerable. For a depth of 8 feet or less approximately the following will be needed: two masons, four helpers; four masons, seven helpers; eight masons, fourteen helpers.

Besides the teams employed in hauling material to the work there should be one for carrying from place to place mortar-boxes, tool-boxes, and other heavy articles.

It is difficult to say anything definite concerning the number of men which should form an excavating-gang. There should be sufficient to keep the pipe-layers or masons constantly at work. Each gang or set of gangs to which a pair of pipe-layers or force of masons is assigned should be just large enough to open and back-fill trench at the average rate at which the sewer is laid. If the sewer frequently varies in depth or ease of digging it is often well to assign a force of masons or pipe-layers to two gangs, always endeavoring to so arrange that one of these is in soil rapidly trenched whenever the other is in deep or difficult work. For 8-foot excavation in good soil requiring little bracing 25 to 30 men at the shovel is usually an economical number; at 15 feet, if no excavating-machinery is used, 60 to 80 will be required for equally rapid work. On account of the considerable sheathing necessary at such depths and for other reasons it may be better, however, to still maintain the gang at 25 or 30 men, and assign the sewer-masons or -layers to two gangs. It is usually undesirable to change the size or personnel of gangs after they have once been gotten into good working shape.

If a trench runs into very wet soil or quick or running sand gangs as large as the above cannot be used to advantage, since not only must sheathing be set and driven right up to the excavation as it proceeds, but the pipe or sub-drain must be laid or foundation put in foot by foot as the bottom of the trench is reached; also an upheaval of the quicksand bottom, caving, and other accidents may cause occasional stoppages of the work for a few minutes, when almost the entire gang must lie idle or go to back-filling. In such difficult work on pipe sewers a gang may consist of a foreman, a sheather, two pipe-layers, and five or six laborers. If the ground is very wet it is advisable to open only a little trench at a time, since the more that is open below water-level the greater the amount of water which will flow through the trench and interfere with the work. Under such circumstances the gangs should be small.

If the back-filling is not to be rammed it is the custom of many contractors to use the entire gang for the last 20 to 30 minutes each day in back-filling. This arrangement has the advantage of not requiring an extra gang and foreman for back-filling. But if there are three or more gangs excavating it would perhaps be better to keep one gang continually back-filling. This is certainly advisable in all cases where the trench is to be thoroughly tamped.

The contractor, general foreman, or timekeeper should visit each gang just after the beginning and just before the ending of each day's work, at the least, to learn of any material needed or difficulty encountered, and also to get the "time" of the men, which may have been taken by the foreman, or may better be taken directly by one of the three above mentioned.

If Italians or other non-resident workmen be employed (and if the work is in a small city and requires many men outside labor must be obtained) they are usually housed together in barns or empty houses or shanties constructed for the purpose on the outskirts of the city. If these can be located near a stream the men will usually take advantage of the opportunity to wash themselves and their clothes and keep in better health than if otherwise situated. The necessity for walking a long distance to and from work will result in decreased energy in their labor, and should be avoided. It will sometimes pay to have the teams carry them to and from the work. It will also be to the contractor's advantage to see that their food is wholesome. A considerable experience with Italian laborers has convinced the author that as a class they are more appreciative than are native laborers of both kindness and harsh treatment, and are shrewd readers of motives of conduct. If justly though firmly treated they are polite, obedient, and good workers, slow to wrath, but dangerous if ill treated. "Sore-heads" among them should be gotten rid of at once.

Pay-days should come at as long intervals as possible, because of the diminished force which can be made to work for the following day or two, if for no other reason. For some reason masons seem to be peculiarly subject to the failing of "pay-day drunks," and if possible an arrangement should be made with them to pay their expenses wherever they wish to board and a small weekly amount of pocket-money, the balance being paid them when their work is completed. Monthly payments are generally made to the laborers, immediately after the payment of the monthly estimates.

ART. 70. TRENCHING BY HAND.

The line of the trench being given by centre stakes, the sides of the excavation are indicated by measuring the proper distance on each side of the stakes and stretching sash-cord or clothes-line there and marking the ground along this line by means of a pick. The laborers are then placed at regular intervals along the trench, varying from 6 to 20 feet, in single line in most cases, but if the trench is 8 feet or more wide they may be in double line. It may be well to define in some way, as by a mark in the ground or stake at one side of the trench, equal lengths of trench, one man being required to work within the limits of each length. Where possible it is desirable that this length be that which can be completed in a half or a whole day.

If the street is macadamized or gravelled or has a hard dirt surface a contractor's "rooter plow" may be used to break the surface; but this is not advisable in narrow trenches, nor should the surface be broken beyond the sides of the trench, since if sound it helps to prevent caving of the sides.

If there is any paving material on the street it should be thrown upon one side of the trench, and the remaining excavated material upon the other side, the material on each side being kept back a foot or two from the edge of the trench to allow a pathway for foremen and inspector and for lowering material, but still more to prevent excavated material from falling back into the trench. Thus one side of the street is left open to travel, the pile of paving material acting as a guard to the trench on that side. If so much soil is to be thrown out or the street is so narrow that it cannot all be placed upon one side of the trench it may be placed upon both sides, the paving material being kept separate, say along the outside edge of one bank; but it would be better to use excavating-machinery and thus avoid blocking the street entirely. The amount which can be placed upon one side of the street without covering the sidewalk may be increased by setting there a platform and guard, as shown in Fig. 11.

The first earth cast out should be thrown to what will be the outside edge of the bank, since it cannot be thrown there when the trench is deeper without double handling. The gutters should be kept open and free from any excavated material. Down to a depth of 9 to 12 feet the earth can be cast to the surface, although after 5 or 6 feet is reached it will

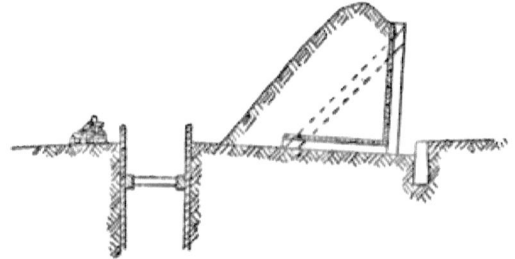

Fig. 11.—Excavation-platform.

be necessary to keep additional men on the surface to throw back onto the pile the material so cast out. When the depth exceeds 9 to 12 feet it will be necessary to handle the material twice before it reaches the surface, by placing a platform or staging about 6 or 7 feet below the surface, onto which the earth is thrown by two to four men, and from which it is thrown to the surface by one man. If the depth exceeds 16 or 18 feet still another platform will be necessary about 7 or 8 feet below the first. These platforms are usually made by resting plank upon the braces or rangers of the sheathing. (Except in rock cuts there are almost no conditions under which a trench 10 feet or more in depth should be left unbraced.) The platform may consist of short pieces of plank placed crosswise of the trench, their ends resting on the rangers, or of long plank lengthwise of the trench resting upon the braces. The latter cannot well be used if the trench is less than 5 feet wide, but is the better form for wide trenches. If there is more than one tier of longitudinal platforms the successive tiers should be placed alternately upon opposite sides of the trench; or if cross-platforms are used the right side of one should be vertically above the left side of the next lower, alternate platforms being vertically above each other. The number of men excavating which cast onto one platform

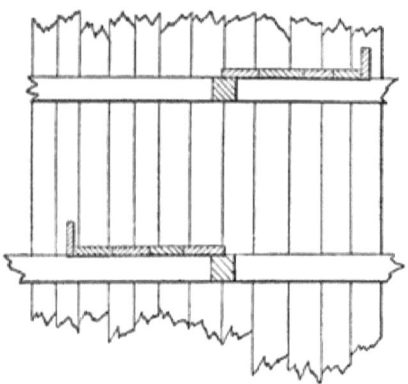

FIG. 12.—CROSS-STAGING IN TRENCH.

may be only two, but should increase with the difficulty of excavating, so as to keep the staging-man busy.

Where it is allowed (as it is in many cities), and the trench is over 10 feet deep, it is often economical, except in hard rock, dry sand, or quicksand, to make the excavation in alternate tunnels and open trenching, the sections of each being 8 to 20 feet long. The tunnel is usually made about 5 feet high. The amount of material to be removed and of bracing to be put in is thus reduced. But tunnelling should never be allowed under streets, except in rock, unless the tunnel is afterwards opened and back-filled as open trench, being used only to save bracing; since it is practically impossible to so compact the back-filling in a tunnel as to prevent future settlement, which may not occur, however, until months or years later, when the contractor has been relieved of all responsibility. Where the amount of traffic on a street or other conditions require it, however, a tunnel may be run under the street and a masonry lining, which may be the sewer itself, built against the outside of the excavation, so that there is no back-filling except in the form of masonry; which construction requires special tunnelling-machinery and methods. In Paris by the use of a shield a tunnel 19 feet outside diameter was run with a covering in some places of only 2 feet between it and the street-paving above, without causing any cracks in the latter. The successful tunnelling for the Boston underground railway is familiar to all. A notable instance of sewer-tunnelling is found in the sewers tunnelled through sand-rock at St. Paul, Minn., the tunnel, when lined on the bottom, constituting the sewer. Restrictions against tunnelling should not of course apply to lines whose depth is 75 feet or more, such as those passing through ridges.

There is a tendency, if a right-handed laborer always faces one way while picking, for the trench to work to his left as it descends. He should be taught to avoid this by keeping his left side to the side of the trench at which he is picking, so that both sides shall make the same angle, if any, with the vertical.

It pays to keep the picks sharpened and good shovels in the men's hands. For this purpose there should be 25 to 100 per cent more picks than laborers, to allow opportunity for sharpening them. For digging the round-pointed shovel is best, but staging-men and mortar-mixers should use square-pointed shovels. There should be a few extra shovels constantly on hand, including a few long-handled ones, but these latter should not be used for trenching except in deep trenches where the shovelling is very easy.

In soil where caving is frequent and sheathing is not used the trench should be refilled as soon as possible, since the longer it stands the greater the probability of caving. Soils, such as clay or other heavy ground, having some cohesion will usually give warning of caving by cracking a few feet back from the edge of the trench, and should be braced as soon as such sign appears. Gravelly soils or dry sand usually give no such warning, and are particularly dangerous on this account and because they may bury and suffocate the men; while clay, coming in lumps, although it may bury and even crush them, will permit them to breathe until they can be rescued. Trenches in gravelly and sandy soil should always be sheathed.

If a boulder is met with it may be raised from the trench by a derrick or, if too large for this, may be blasted. Before blasting the earth should be removed from all sides of the boulder and the trench in the vicinity should be braced. It may sometimes be cheaper to dig a hole in the side of the bank and roll the boulder into this out of the way. In some cases, when the sewer would pass entirely under the boulder, this may be left undisturbed and tunnelled under. If it merely protrudes into the trench a portion may be removed by "feathers and wedge" or a heavy sledge.

If a water- or gas-pipe or other conduit run diagonally across a trench, or run in it, or cross one more than 8 or 10 feet wide, it should be supported in position before the earth is removed from under it. This can be done by placing across the trench at intervals of 12 feet sufficiently strong timbers or old rails, and suspending the conduit from these by chains drawn tight by driving wedges between them and the beams. Rope should not be used for this purpose, as rain causes it to contract or break in the attempt to do so. If such a pipe lies in the bank, close to or slightly protruding into the trench, the bank should be thoroughly braced just under the pipe and the pipe itself be held in place by braces. These braces should not be removed when the trench is backfilled, and if the pipe is suspended the trench should be filled and thoroughly tamped under and around the pipe before the chains are removed. The breaking of a water-main in or near a sewer-trench is one of the most disastrous accidents which can happen to it. Small house-connection pipes crossing the trench are apt to be broken by workmen climbing over them and should be protected, as by a piece of plank or of a 2 × 4 placed across the trench just above such pipe, the ends extending 6 inches or more into the banks for support. In all cases where there is danger from water-pipe such and so many gates should be temporarily closed that the closing of only one more will entirely shut off the pressure from the threatening line of pipe, and a wrench be kept at hand for closing this.

If a drain crosses the trench the pipe should be removed and saved, and a trough substituted during construction, its ends supported in the banks. The back-filling should be carefully tamped under this and the pipe relaid in the trough.

At the first sign of quicksand the best of close sheathing should be at once put in, an experienced foreman put over this work and the best men placed upon it (see Art. 72).

The soil where a trench has previously been dug, although it were years before, is more liable to cave than that which has never been disturbed, and the sewer-trench should be kept several feet from such old trench if possible.

Art. 71. Excavating-machinery.

As a general statement it may be said that it does not pay to use any kind of machinery in excavating where the trench is less than 8 or 9 feet deep or wide, although it may be desirable or necessary to do so where for some reason the excavated material cannot be piled along the side of the trench. The advantages attending the use of machinery are: greater amount of material excavated with a given number of men, less danger of caving of banks from the weight of earth piled upon them, less obstruction to street traffic, the convenience of having at hand means for raising boulders, lowering heavy pipes, or other material. Each of these advantages increases in force with the depth of the sewer. With several of the machines now on the market the cost of handling material increases but little with the depth. The machinery in use varies from an ordinary boom-derrick to an elaborate system of trestles, wire ropes, and buckets, which may stretch along 400 feet or more of trench.

For a large brick sewer a handy arrangement is that of two derricks with booms about 40 feet long, both placed on the same side of the trench and about 75 feet apart. Both boom- and main-falls should wind upon drums driven by steam-power. With this arrangement a bucket of earth can be hoisted from the excavation and, passing from one derrick to another, be deposited in the trench 125 or even 145 feet away. This plan is not adapted to narrow trenches nor to those where any considerable length of trench is to be under excavation at one time. For these one of the trestle-machines or cable-ways is preferable, the former more particularly for trenches up to 12 or 15 feet in width, the latter for wider ones and for particular cases, such as crossings of railroad-tracks.

The cable-way consists essentially of a wire cable suspended over the centre of the trench, on which run one or more travellers carrying buckets; the earth being excavated at one point and cast into the buckets, which are raised and carried to the other end of the cable, where they dump the earth upon the completed sewer. It is essential to the safety of the laborers that the cable be most substantially anchored at the ends, and that it be amply strong to carry any load which can possibly come upon it. The anchorage is usually in the shape of a "dead-man," but the ordinary log placed

PLATE XII.—TRESTLE EXCAVATING-MACHINE AT WORK.

in the trench and covered with earth back-filling should not be relied upon. Rock may be piled in front of and over the log, but a better plan is to bury in the trench a platform of stout timber, inclined backward about 45° from the vertical, to which the cable is fastened. The hoisting- and conveying-ropes are driven by an engine located at one end of the cable. Like derricks, the cable-way is not adapted to trenches which move forward rapidly, as the moving and resetting of it take considerable time and labor.

In the trestle-machine the buckets travel upon an overhead track which is supported at intervals by trestles spanning the trench. Generally from 6 to 20 buckets are in use at once, one half of which are being filled while the remainder are being carried to the dump and emptied. In some machines the track forms a long loop, one side of which is for going and the other for returning buckets. There are then three sets of buckets, one going to the dump, one returning, and one set being filled. To obtain the greatest efficiency of the machine the number of men casting into each bucket should be just sufficient to fill it during the time occupied in removing, emptying, and returning a set of buckets.

Such machinery is economical when the cost of running—including all labor but that of the men digging in the trench—and of repairs, plus the rental or interest on first cost of the machine, is less than the cost of " staging " it out (as the use of platforms is called) plus that of back-filling. If the back-filling is to be hand-tamped this last item should not be included, since if a machine is used the material must be spread after dumping. A good trench-machine is usually economical when either depth or breadth of trench exceeds 8 to 10 feet in ground capable of rapid trenching; but this economical least dimension increases with the decrease in the rapidity of excavation possible. Where for some reason the excavation can proceed but slowly the use of machinery is not advisable for economical, though it may be for other, reasons.

Whatever the machinery employed it should work successfully although the sheathing extend at least 6 feet above the surface along each side of the trench, should be able to drop a bucket anywhere in the trench, each bucket being always under perfect control, and no cable or rope should hang within 6 feet of the ground. It is better that it should have no cross-ties or other parts extending across the trench within 6 feet of the ground, and that it should not obstruct the street for more than 2 feet on each side of the trench.

For deep trenching through city streets the use of excavating-machinery is strongly recommended as of advantage to both city and contractor.

Most makes of excavating-machinery can be either rented or bought. For a village or small city the former is generally preferable if the work on which it is to be used can be pushed. But if it will be needed for more than one season it may be preferable to buy instead.

Probably the best-known and most extensively used trenching-machinery are the Carson, of Boston, Mass., and the Moore, of Buffalo, N. Y. Other machines are described in Engineering Record, vol. XVI, page 123, vol. XXXIII, page 100; and in Engineering News, vol. XXIV, page 268, vol. XXV, page 547, vol. XXXVII, page 50.

Art. 72. Sheathing.

Just when a trench can be relied upon to stand without sheathing and when it cannot is something that only experience can teach. Sheathing is expensive, but not so expensive as excavating a trench which has begun to cave, to say nothing of settling for injuries and death of laborers. If earth has been piled upon a bank which afterwards caves it may be necessary to re-excavate more material than all that which would have been excavated had no caving occurred, and all of this must be removed to some distance because there is no bank upon which to pile it. Not only that, but the soil is liable to continue to slide into the trench, making it almost impossible to keep the bottom uncovered. If, after caving has begun, sheathing is used the difficulty of placing it is greatly increased. A trench which if sheathed would have given no trouble may become a most discouraging hole into which many times the cost of sheathing must be placed in the form of labor before the sewer is built therein. The author's experience has been that it does not pay to take many chances with unsheathed trenches. He would use at least skeleton sheathing in every trench more than 8 or 10 feet deep, in any trench in gravelly and sandy soil, and whenever the least sign of caving appears. Wherever the street is paved a plank should be placed horizontally on each side of the trench about 6 inches below the surface, and braces driven between these not more than 6 feet apart.

Sheathing is usually placed as follows: A plank (a, Fig. 13) is placed upright in the trench against the bank, another (b) 12 feet from this, and two against the other bank and directly opposite these. Against each two and near the street-surface is placed a horizontal ranger (cd and ef), both at the same level, and between them at each end a brace is driven, long enough to be a tight fit. Two other rangers (gh and kl) are placed, one on each side of the trench, from 4 to 6 feet below the others, and braced. Sometimes these lower rangers are placed first. The ends of the rangers come in the middle of the uprights, the braces only an inch or two from the ends of the rangers. The next set of rangers abut against these and are braced in the same way. Generally an additional upright and braces are placed midway of each ranger. This forms skeleton sheathing.

If the sheathing is to be close, plank are slipped behind the rangers and in contact with each other, and one or more additional braces are placed at equal intervals between each tier of rangers. For bracing only, the rangers and braces are used without any vertical sheathing. These are ordinarily placed a foot or two from the surface, or just beneath and in

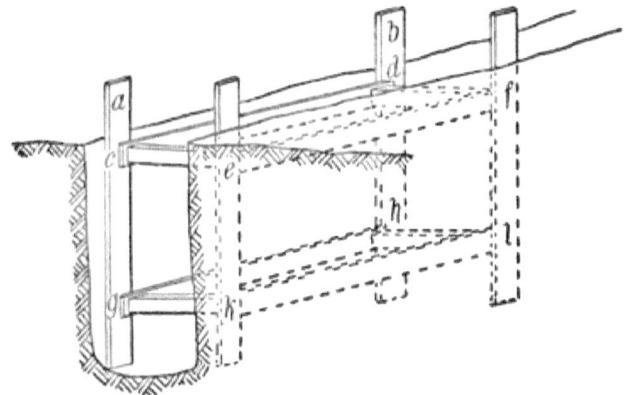

FIG. 13.—SKELETON SHEATHING.

front of an exposed water-main or other conduit. When a series of rangers and braces are placed one just below the other horizontal sheathing is formed.

As the trench is deepened the sheathing should be driven so that its lower end is as near as possible to the bottom of the trench, unless rock or some firm soil be previously reached. *In quicksand or running sand the bottom of the sheathing should always be kept at least one foot below the bottom of the excavation.* This is essential if the work is to be done without considerable loss of money and perhaps of life. As many men as are necessary to insure this should be kept constantly at work driving the sheathing. No two planks behind the same ranger should be driven at once, as the latter would in that case be apt to follow them down, which it should not do.

If there is a tendency for the sheathing to be forced in at

the bottom by the bank, as in the case of quick or running sand, a set of rangers and braces should be put in place immediately under the lowest set already in position as soon as the excavation is low enough to permit it. As the excavation and sheathing are carried down this last set of rangers should be driven down, always being kept level crosswise of the trench and just above its bottom, until it is the proper distance below the preceding set, when it is driven no further, but another set is started under it. If the bank is tolerably stable the stiffness of the sheathing-plank can be relied upon to keep it in place below the second ranger until the trench is sufficiently deep to permit placing the third ranger in its proper position without any further driving.

Each brace should be exactly beneath the braces in the tiers above.

There is considerable friction between the sheathing and the bank on one side and rangers on the other, and after two sets of rangers are in the driving becomes quite difficult and the upper ends of the plank become battered and broomed and the plank broken, sometimes even when they are protected by caps. Ordinarily 10 or 12 feet is the greatest depth to which plank can be driven economically. It then becomes necessary to start a new course of sheathing, which is placed inside the upper course, its back resting against the rangers of this. The second course is driven and held in place by rangers and braces as was the other, and may be succeeded by one or more other courses each 10 or 12 feet high. When a new course of sheathing is started it is advisable to temporarily fasten planks horizontally in front of and behind this sheathing near the top, by a nail at each end driven into a sheathing-plank, to keep the plank in line and steady them while driving.

In placing each course after the first one an opening must be left at each vertical line of braces, since the sheathing cannot be driven there. If these openings give trouble they may be closed by slipping into them, behind the rangers, 5-foot lengths of plank when the trench has reached that distance below the first course of sheathing, and driving these to keep pace with the other sheathing. When the trench is 5 feet deeper still another 5-foot length of plank may be slipped behind the ranger on top of the former length, and so on. A short piece of plank, at least, should be kept in the bottom of this opening to keep the planks on either side the proper distance apart.

Another method of closing these openings is to cut a plank just long enough to reach from the bottom of one ranger to

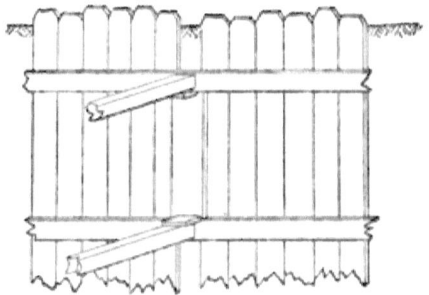

FIG. 14.—SHEATHING UNDER BRACES.

the top of the next and a little wider than the opening. This is placed over the opening against the face of the sheathing, and between the rangers, to which blocks are nailed to hold it in this position (see Fig. 14).

In some cases it will not do to leave this space open for even a foot above the bottom of the trench, as in quicksand. It may then be advisable to use a somewhat different system of rangers, as follows: In placing rangers for the first course of sheathing, where one ranger is ordinarily placed two will be placed, one in front of the other but separated from it by a small piece of plank at the end of each brace. The front ranger may be but a 2-inch plank. The second course of sheathing is slipped between the two rangers and when it is all in place except where the spacing-blocks interfere the braces are driven along about a foot, the spacing-blocks knocked out, and sheathing dropped into the spaces they occupied. Generally plank behind a brace cannot be driven, owing to the friction, but when the one next to it has been driven the brace can be moved over in front of this and the former then driven.

Where there is more than one course of sheathing, or whenever the bottom of any course is not kept at the bottom of the trench, all braces in each vertical line should be tied together by cross-bracing of plank nailed to them; otherwise one side of the sheathing may drop, loosening the braces and causing a complete collapse of sheathing and trench. The author has seen several serious accidents due to the neglect to use such cross-bracing.

The sheathing is usually of hemlock plank, although pine would be better, being less brittle. Maple and other hard wood has been used in a few instances. The plank is usually 2 inches thick, although heavier may be advisable in deep, wide trenches or where it is desirable to use as few rangers and braces as possible. It should never be less than 2 inches thick. Ten or 12 feet is the usual length, although 18 or more is sometimes used. But the great amount of friction between such long plank and the earth makes it extremely difficult to drive the last 6 or 8 feet, the top of the plank being usually broomed or broken in the attempt. For the same reason the width of the plank does not usually exceed 6 or 8 inches. All the sheathing in a given course should be of approximately the same length. Sheathing-plank should be sharpened to a chisel edge, the flat side being placed against the bank, and the edge which will not be in contact with the plank last driven should be bevelled, that the plank may hug the bank and keep a close joint with the one previously driven. The bevel may be 3 to 5 inches long. The top of the sheathing-plank should be bevelled on each edge, to lessen splitting and binding and to permit of using a driving-cap, which is advisable if the sheathing drives hard, to keep the plank from brooming.

For driving the sheathing a hardwood maul is ordinarily

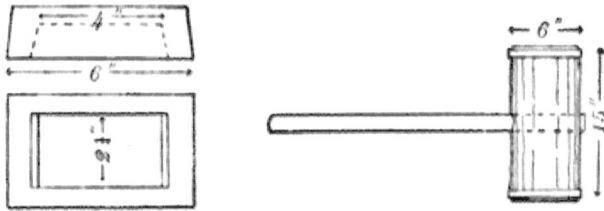

FIG. 15.—DRIVING-CAP AND MAUL.

used, about 6 inches in diameter and 15 inches long, with a wrought-iron hoop banding each end.

If a large amount of sheathing is to be driven in deep trenches a steam-hammer pile-driver may be used to advantage. This does not broom the pile, and by using it sheathing 18 feet long or more may be driven. It is particularly applicable to sand and elastic soils.

If the ground is such as to require sheathing from the

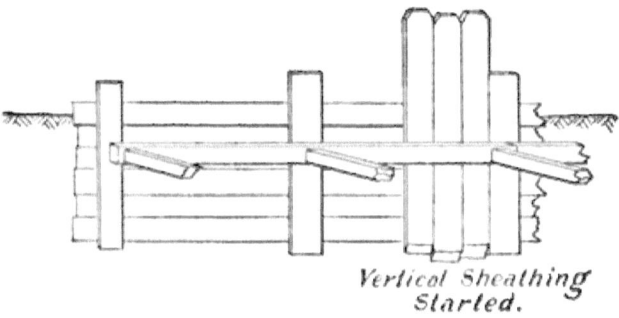

FIG. 16.—HORIZONTAL SHEATHING.

very beginning of the excavation it would be difficult to keep vertical sheathing standing and in line while the trench is only 1 to 3 feet deep, and it would greatly interfere with casting out the excavated material. It will be better in such a case to erect skeleton sheathing, with only one set of rangers and braces and short uprights, behind the uprights placing plank laid horizontally. When this construction has been carried down 5 or 6 feet vertical sheathing can be started and continued as above. But even then if the vertical sheathing is more than 8 feet long it will be necessary to use platforms or staging, unless a sheathing-plank can be omitted every 5 or 6 feet and the earth cast out through the opening thus left. On account of the difficulties just described it is better, if the trench is so deep as to require more than one course of sheathing, to place shorter sheathing in the top course—for instance, 6-foot sheathing and then 12-foot in a 15- to 18-foot trench.

Some contractors use horizontal sheathing altogether, the verticals being only 3 or 4 feet long, several being placed one above the other. Most American contractors, however, prefer the vertical sheathing.

The size of the rangers may vary between wide limits, but in any one trench they should all be the same length, and when in position the ends of all should come opposite or under each other. Two-inch plank may be used for rangers in ordinary loamy or clayey soil and shallow trenches, and the braces placed with sufficient frequency to prevent their bellying too much. This would in many cases bring the braces so close to each other as to interfere with the work, and it will then be advisable to use 4×4 or 4×6 material. The author prefers these in any case, as being stronger, but neither costing nor weighing more, than 2-inch plank. If excavating-machinery is used the braces should be at least 5 or 6 feet apart, and the rangers of 4×6 or 6×8 timber. The deeper the trench the heavier should be the rangers and braces.

The braces should be heavier also the wider the trench,

since they must act as posts. They are often, for convenience, made of the same size of timber as that used for rangers. Each brace must be fitted to its place, since the width of a trench usually varies at different points within a range of several inches. For finding the length of brace

FIG. 17.—SLIDING ROD FOR MEASURING BRACES.

required it is handier to use a sliding rod than a measuring-rule. The brace should be made a little longer than the distance between rangers, that it may drive hard into place and fit there tightly. To make this driving easier one edge of one end of the brace may be slightly bevelled.

Instead of wooden braces extensible iron ones are coming into general use, and for narrow trenches at least are equally as good and much more convenient, since they can be quickly adjusted to any position and used over and over again. For wide trenches those in the market are hardly stiff enough, but are apt to buckle under extreme pressure. Trussed beams, however, can be obtained with extensible ends, which meet this objection. If much bracing is to be done the cost of extensible braces can be saved in the carpenters' wages many times over.

Much heavier sheathing than here described may be necessary in deep trenches in some soils. In stiff marsh-land near New York City, in a trench 26 feet wide and 25 feet deep, 6-inch sheathing was found necessary, with 10×10 rangers and 8×8 braces 5 feet apart horizontally and vertically.

In cases where the soil was soft round piles have been driven a few feet apart along the side lines of the trench before excavation, and as this proceeded horizontal sheathing was inserted behind the piles and braces placed across the trench between them.

The rangers and braces can be used over and over again if they are not left in the trench; the sheathing, too, can ordinarily be used several times; but each time a set is used a few plank will probably be broken, either in driving or in drawing. As stated in connection with Table No. 21, good sheathing can ordinarily be used two to five times, taking an average of all used at the outset.

In many instances it is desirable to leave the sheathing in the trench, sometimes with and sometimes without the rangers and braces. The conditions calling for leaving in sheathing are: that drawing it may endanger the sewer, or water- or gas-pipes in the street near the trench, or adjacent buildings, or that the street-paving will be injured thereby. The danger to buildings usually exists only in connection with deep trenches in unstable soil or where a building is quite near a sewer which lies below its foundation. Water- or gas-mains would be endangered if within two or three feet of, and more than that distance above the bottom of, a sewer-trench in fairly good soil. If the soil has shown a tendency to crack along the banks near the trench the sheathing should not be drawn if the street is well paved; and if water- or gas-pipe or other sewers are laid in such street the judgment of the engineer must decide at what distance they may be considered safe from disturbance if the sheathing be drawn. If the sheathing has been driven below the centre of a sewer, as must be done under some conditions, its removal would disturb the foundation of the sewer and should not be attempted. But if two or more courses of sheathing have been driven all but the lowest course may be removed if the sewer only is affected. The rangers and braces as well as the plank should usually be left in. If the banks are liable to cave with the drawing of the sheathing the trench should be filled to a distance above the sewer at least equal to its width before the top braces are knocked out or any sheathing-plank is entirely drawn.

Before drawing sheathing the back-filling, if it is not to be rammed, should be carried to a point at least 3 feet above the bottom of the plank. The bottom set of braces and rangers may then be removed. If this gives less than 2 feet of back-filling above the top of the sewer this amount should be thrown in and properly tamped. When the sewer is properly covered the remaining braces and rangers may be removed and the sheathing entirely drawn. If the bank should cave badly on the removal of the braces it might break the sheathing, and in such a case it may be better to continue back-filling and slowly drawing the sheathing, each set of rangers being removed only as the back-filling reaches them. If there is more than one course of sheathing this plan should be followed in every case with all but the top course, unless the others are to be left in the trench, which may be cheaper in some cases.

Drawing the sheathing is often a difficult matter if only the hands or a pick be used. A convenient plan is to use a sheathing-puller, made of iron $1\frac{1}{2}$ or 2 inches thick and 3 or

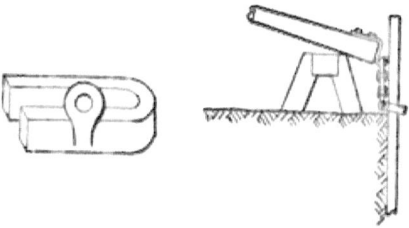

FIG. 18.—SHEATHING-PULLER.

4 inches wide. The ring on the clamp should be so placed that the clamp will slide down the sheathing when not supported, remaining constantly horizontal. After placing this in position on a horse a simple pump-handle motion with the lever will suffice to draw the plank. A chain to be hooked tightly around a sheathing-plank may be used as a substitute for the clamp, but is not convenient for close

sheathing, which must be pried apart to admit it. Better than this sheathing puller, where excavating-machinery is being used, is to use the engine-power to draw the sheathing by fastening the clamp to a hoisting-rope.

Where a building is so situated with reference to the sewer-trench that its stability is endangered thereby the greatest care should be taken with the sheathing to prevent any material behind it from caving into or in any way entering the trench. To insure this the sheathing-plank must be tight together—in sand it may be necessary to use tongued and grooved plank—and their bottoms should be kept well below the bottom of the trench. If this is done and the bracing is strong and stiff there should be little danger, unless the material is semi-fluid, when it may be impossible to prevent a settlement of the ground and buildings, unless by freezing the soil by the Poetsch process (an exceedingly expensive one) or some similar method.

If a settlement of a portion of a building-foundation seems probable the building should be shored and jacked up. One method of accomplishing this is to make openings just above the ground-surface 6 to 10 feet apart and of a size to permit large beams—10×12, or 12×14 or 18—to be passed through them. These beams are supported at each end by jacks, which in turn rest upon blocking placed upon the ground. A careful watch is kept of these and at the least sign of settlement of the ground the jack above is screwed up an amount equal to this settlement. As a further precaution it may be advisable to shore up the walls by a sufficient number of heavy timbers, whose lower ends are supported upon platforms or grillage, wedges being placed under the foot of each and driven up when necessary to make up any settlement of the ground. The shores at their upper ends bear against beams bolted to the walls of the building, or in masonry walls are received by openings about a foot deep cut therein. Shores alone are often employed when the building is not valuable or the danger is small.

ART. 73. LAYING SEWER-PIPE.

It will save considerable trouble in the laying of pipe if the foreman has the trench dug exactly to line and grade as ascertained by measuring and plumbing from a grade-line already set. It is better to have the bottom a little too high rather than too low.

Pipe sewer is usually laid up the grade, and the pipes are so manufactured that the specials must be laid with their bell ends pointing up. Laying the sewer-pipe in this way is more likely to produce good joints, particularly if the grade is at all steep, since if laid down grade a pipe, after being placed in position and before the next is laid, tends to slide away from the one next above it and cause a break in the inner surface of the sewer and a leaky joint. It is also much easier to lay pipe with the bell pointing ahead, and the cement joint is apt to be firmer. The only reason advanced for laying pipe down hill is that the lower end of the trench being ahead of the pipe, any ground-water will be kept drained away from the sewer construction. This is discussed in Art. 78.

For lowering into the trench pipe which does not weigh more than 100 pounds a convenient method is to use a rope of $\frac{3}{4}$ to $1\frac{1}{4}$ inch diameter with a hook at one end. The hook is passed through the pipe from spigot to bell and then back over the outside to the middle of the pipe and caught on the rope there, so that the pipe when suspended will be horizontal. Or the hook may pass through the pipe from bell to spigot and be simply caught over the end of the latter. The pipe is lowered over the edge of the trench by one man and received at the bottom by another if light, or by two if heavy, the hook being unfastened and pulled up. If the pipe weighs more than 100 pounds two men will be required to lower it, which they do by each holding one end of a rope which passes through it. For pipe heavier than 200 pounds it is advisable to use an ordinary three-leg derrick with light tackle-block. The pipe is then suspended by a rope or chain, with a hook at one end and a ring at the other, passed through the pipe and so hooked that it may be lowered in a horizontal position. A convenient arrangement for holding the pipe consists of a hook (Fig. 19), which should be at least two thirds the length of the pipe and very strong at the bend. The ring must come beyond the centre of gravity of the pipe to prevent its falling off the hook. By use of this a pipe can while suspended be entered into the bell of the one previously laid and much heavy lifting by hand avoided.

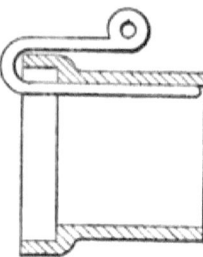

FIG. 19.—PIPE-LAYING HOOK.

Another method of entering heavy pipe after it is in the trench is sufficiently explained by the illustration Fig. 20.

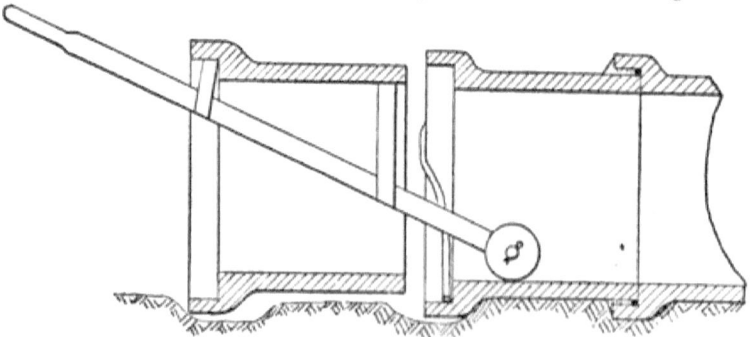

FIG. 20.—APPLIANCE FOR "ENTERING" HEAVY PIPE.

This is made of wood or iron, with a loose wheel on either side of the bar at the bottom.

Before a pipe is lowered into the trench a "bell-hole" should be dug where its bell will come, of such size that when the pipe is in position the jointer can pass his hand entirely under and around the front of the bell. It is convenient to have a stick exactly as long as two or three lengths of pipe, by which the location of each bell-hole is measured from pipe already laid, the bell-holes being dug for a few lengths in advance of the sewer.

Two men should be employed in laying sewer-pipe, one straddling the pipe last laid, the other in the trench just ahead of it. The latter as the pipe is lowered guides it into place and releases the hook on the lowering-rope, if one is used. The former, holding one end of a length of packing in each hand, places the loop thus formed under and around the pipe about an inch from the spigot end and guides this into the bell of the pipe last laid, taking care that the packing also enters the bell. With a yarning-iron he then pushes the packing up against the shoulder of the bell all around, being first sure that the pipe is " home " in the bell. The other pipe-layer meantime supports the pipe at the bell end and shoves it home. The grade-rod and plumb-bob are then used. If the bell end is too high (the spigot end should be all right, since the previous pipe is) it may, if the soil is loam or loose clay or sand, be forced down a quarter of an inch, more or less, by standing and jumping upon the top of the pipe. (The pipe-layer should never rest his foot *inside* the pipe to force it down, as this is likely to break the bell or even the pipe.) If the soil is stiff clay or gravel the pipe should be removed and the trench bottom lowered sufficiently with the shovel. If the pipe is too low it should not be raised by placing a stone or piece of wood under it, but should be removed and fine earth placed and rammed in the bottom of the trench. By means of the plumb-bob the pipe should be centred exactly under the grade-line. A convenient way of doing this is to suspend the bob from the cord at a grade-plank, being careful not to lower the cord by its weight; then, when the eye is so placed that the cord and plumb-bob string coincide, the former is projected by the eye vertically into the trench and should cut the centre of the pipe. With a circular salt-glazed pipe the centre is known by a streak of light reflected from the sky, and this streak should be bisected by the vertical projection of the grade-cord. Another plan for obtaining a vertical projection of the grade-cord is to stretch another cord a foot or two vertically below it. But this method is less accurate in practice than the other and is not recommended. The grade-cord cannot be stretched so tight that it will not sag $\frac{1}{18}$ to $\frac{1}{4}$ of an inch at the centre, but allowance may be made for this in using the grade-rod. The foreman or inspector who uses the grade-rod will need to have a short movable plank spanning the trench just ahead of the pipe being laid, on which to stand.

As soon as a pipe is in position sufficient earth should be placed and rammed on each side of it just back of the bell to prevent its moving. The next pipe is then lowered and set, and so on.

At least two joints behind the pipe which is being set is another man, who cements the joints. The cement he usually keeps in an iron pail of ordinary size (although one having the shape of a pan would be better), just enough being mixed at a time to permit his using it all before it stiffens. If there is any delay in laying the pipe the pail should be cleaned out lest the cement set in it. The jointer should wear rubber mittens, and a small trowel will be found more convenient than the fingers for getting the cement out of the pail. The cement mortar should ordinarily be about as stiff as putty, but if the trench is wet it should be as dry as it can be and have any cohesion. The jointer takes a handful of mortar in each hand and presses it into the bell all around, drawing his hands meantime around the joint. With a wooden or iron calking-tool he compacts the cement in the joint, adding more as is necessary, and with additional mortar he makes a neat bevel outside the bell, continually pressing the mortar firmly towards the bell. This bevel should not be flatter than 45°, since if too much mortar be outside the bell its weight may cause it to fall away from the pipe and perhaps draw with it the mortar from inside the bell. The compacting of the cement is frequently omitted, but is necessary if tight joints are to be obtained.

Just behind the jointer should be another man, who, as soon as a joint is made, fills the bell-hole carefully with fine earth well tamped, and then fills and tamps the same material under and around the rest of the pipe up to at least its middle. His tamping-bar should be of wood, there being danger of breaking the pipe if the ordinary iron ones are used, and with a face about 2 × 4 inches. If the trench is wet so that water collects in the bell-holes the mortar is likely to become softened and fall out of the joint. To prevent this a piece of cheese-cloth may be wrapped tightly around the joint after it is made, as specified for sub-drains; or the bell-hole may be immediately filled with concrete thoroughly compacted. The latter is the better but more expensive plan. Where there is much water in the trench it is strongly recommended that concrete be placed not only in the bell-holes but entirely around the pipe at the joints (see Art. 46).

In making the joint it is quite probable that some cement will be squeezed into the pipe, forming a ridge or lumps on the inside. To remove these a bag or disk should be drawn through the pipe past the joint as soon as it is finished, which is done by the pipe-layers. The bag may be an ordinary cement or similar sack, somewhat larger than the sewer, filled with straw or excelsior and a rope tied around its mouth and carried through the sewer, being passed through each pipe as it is laid. The bag should fit snugly against the pipe all around. Instead of the bag a disk of heavy rubber packing bolted between two smaller

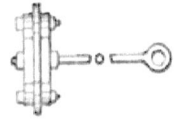

FIG. 21.—PIPE-CLEANING DISK.

wooden disks and fastened to an iron rod may be used, being drawn forward as in the case of the bag. The rubber disk should be slightly larger than the sewer.

When a manhole or other break in the sewer is reached in the pipe-laying the last pipe before reaching and the first after leaving it should be omitted or left with uncemented joint, to be laid while the manhole or other appurtenance is being built. This is on account of the probability of such pipe being disturbed or broken during the construction of the masonry before it has been walled in. In this or any case where a stretch of pipe ends, or when the laying is temporarily stopped, a plug should be inserted in the end of the last pipe, and a bar or stake driven against it into the ground or nailed to the sheathing to hold it in position. The last joint should be left uncemented until laying is renewed.

In setting branch specials the earth where the special will come should be so excavated as to permit the branch to rest upon it firmly when in the desired position. If necessary earth should be placed and tamped under the branch for this purpose. The inspector must not forget to examine each branch to see that a cover is cemented in it, unless the house-connection is to be built at once, and also to mark its location. In wet soil particularly uncovered branches may give rise to serious difficulty, and an unlocated branch is worse than none at all.

If work must be done in the winter-time great care should be taken to prevent the mortar from freezing and to keep ice and frozen dirt out of the joints. In pipe-joints this is not very difficult if the trenches are at all deep, since in these the temperature seldom falls below 40°. But the sand for mortar should be heated, and the pipe also, to insure the removal of all frost from the bells and spigots. In shallow trenches the joints should be covered as soon as possible with at least two feet of unfrozen earth. Care should be taken, particularly when back-filling is dumped from excavator-buckets, that no frozen lumps fall upon the sewer.

The back-filling of trenches has been sufficiently discussed in Art. 54. When this is thrown in without ramming particular care should be taken that all pipe be first well covered with earth, since stones and frozen lumps invariably roll to the foot of the face-slope of the back-filling and might crack unprotected pipe.

It is frequently necessary to cut a sewer-pipe to a certain length or to split one in two to obtain a channel for a manhole bottom. This can be done with a cold-chisel and hammer, a light cut being made first entirely around or along the pipe and this gradually deepened until the pipe of itself breaks in two. The pipe is sometimes filled with sand well packed before the cutting is begun, but this is not necessary if care be used.

ART. 74. BUILDING MASONRY SEWERS.

Circular or egg-shaped masonry sewers may consist of a ring of masonry of uniform thickness throughout, or this ring may be much thicker in the arch than in the invert, or there may be invert-backing masonry resting upon a platform foundation or filling the irregular spaces of a rock cut. If the sewer comes under either of the first two cases it is usually made entirely of either brick or concrete, owing to the expense of dressing stone to make tight work in comparatively thin rings and to give a smooth interior surface. For massive masonry, as in invert-backing or heavy arches, stone can be used and is in many cases cheaper than brick. In some instances concrete may be cheaper and better than either.

A simple ring invert can be used only where the soil is firm and compact enough to stand when given the shape of the outside of the invert; such as clay, pure or mixed with sand or loam. If it will not retain this shape while the sewer is being built, but is solid enough to offer good foundation, as damp sand, the bottom of the trench may be given a flatter curve and lined with a board or plank cradle, upon which concrete or stone masonry is placed for the invert-backing, to be lined with 4 inches of brick-work. In rock cuts the same plan may be adopted, since it is usually impracticable to bring the rock to the exact shape of the sewer (see Plate VI, Fig. 10).

If artificial foundation is necessary this usually consists of a platform, upon which the masonry rests, and which is placed directly upon the trench bottom or supported upon piles.

If the arch is of such dimensions that the thrust is more than the banks can be trusted to sustain, and a shape similar to that shown in Plate VI, Fig. 5 or 9, is adopted, concrete or stone masonry may be used for the side walls, and a platform is generally necessary for foundation except in a rock trench.

Where no invert-backing is necessary the method usually employed is as follows: Templets, two for each gang of masons, are provided conforming to both the inside and outside shape of the sewer. A convenient form is shown in Fig. 22, which is for two rings of brick. This is made of

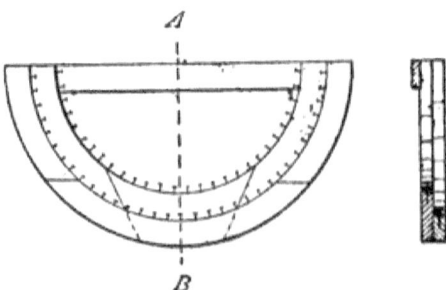

FIG. 22.—TEMPLET FOR BRICK SEWERS.

boards or plank, 2-inch plank being sufficiently heavy for any but very large sewers. A templet for an egg-shaped sewer can of course be made in the same way. Each ring of brick is represented in the templet by a layer of plank, its inside edge conforming to the inner surface of said ring. A number of fourpenny or fivepenny nails are driven along the edge of each plank at equal intervals, the space between them being the thickness of a brick plus that of the mortar-joint, usually about $2\frac{1}{4}$ inches. Each templet should be an exact duplicate of the other, including the position of the nails. At the exact centre A of the cross-piece a notch is cut or a nail driven.

When the bottom of the trench is about to grade one of these templets is set in a vertical position so that the centre of the cross-piece is exactly in the centre line of the sewer, the cross-piece level, and the inside of the templet at the proper grade for the sewer-invert. About 12 to 20 feet along the trench the other templet is similarly set, the sides of the templets containing the nails facing each other. If now a cord is stretched from any nail in one templet to a corresponding nail in the other the excavation should be exactly the same distance outside this as is the outside of the templet. If the excavation should be carried too far it must be filled with sand well rammed, or with good cement mortar.

The cord is now stretched between the lowest nails in the outer rings of the two templets, and the brick laid to this line from end to end. The cord is now shifted to the next nail in the same ring and the next row of brick laid. When two or three courses have been laid on one side of the centre the same number are laid on the other side, and both sides of the sewer are carried up simultaneously, for which reason the masons usually work in gangs of 2, 4, 6, or 8. Not more than the last number can work to advantage on one section of invert, but several sections may be under construction simultaneously.

When four or five courses have been laid a plank is placed on these for the masons to stand on, and the brick-work is continued row by row, each row being laid carefully to line.

The bricks of succeeding courses should break joints at least 3 inches.

After the outer ring has been completed to the springing-line the next is laid in the same way. The bricks of each ring should be bedded in mortar at least $\frac{1}{8}$ inch thick, and every joint should be completely filled. Considerable difficulty will be found in getting any but experienced sewer-masons to lay the brick radially, but smooth work cannot be obtained otherwise and this must be insisted upon. All joints should be carefully struck. If they are not they should be afterward raked out and pointed.

If the brick do not absorb more than 2 or 3 per cent of water in the absorption test they should not be wet, as they cannot then be made to stay in place. But if they take more water than this they should be wet just before using. A quick test for this on the ground is to drop a brick into mortar and remove it. If the mortar does not in two or three minutes grow dry where it touches the brick they probably do not need wetting.

The mortar is usually mixed in a box on the bank (it should never be mixed on the ground for any purpose) and lowered into the trench in a pail by a rope provided with a hook, where it is emptied onto the mortar-boards. These boards are usually 24 to 30 inches square. The brick is placed in hods on the bank and lowered to the masons. A convenient form of hod is shown by Fig. 23, which is made of sheet iron and can be quickly filled and emptied. The material is usually lowered by hand for small sewers, and the man who does this should have a heavy leather palm-piece for each hand. A leather glove or mitten would not last a day of hard usage at such work. To permit of lowering the material a platform is usually thrown across

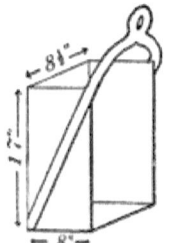

FIG. 23.—HOD FOR LOWERING BRICK.

the trench above where the masons are working. If stone is being laid, or much material is to be used at one place, or the trench is quite deep, the material may be lowered by a windlass set in a portable frame or by a derrick. If excavating-machinery is being used this may be utilized for lowering the material.

As the invert of the sewer rises it becomes difficult for the masons to lay the brick, and the material if in the bottom of the sewer is too far from the work. A platform is then necessary and can be made by sawing plank of such length as to be at the desired elevation when placed horizontally crosswise of the sewer. Three or four of these can be thus laid, with a few brick under the centre of each as additional support,

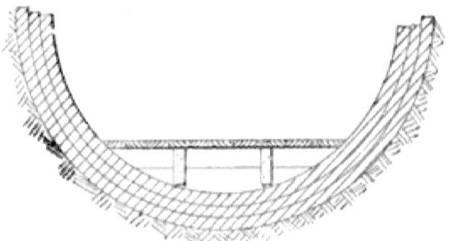

FIG. 24.—MASONS' PLATFORM FOR BRICK SEWERS.

and a platform of loose plank placed over them. But this is apt to distort the green brick-work at the ends of the cross-plank, and it is better to have a number of plank cut to the shape of the sewer-invert cross-section, which will distribute the load along their entire length, and to rest the platform on these (see Fig. 24).

When one section of invert is completed one of the templets is moved ahead the length of a section and set. The other will not be needed by the masons, since one end of the cord will be fastened to nails stuck into the joints of the invert just completed. The second templet can, however, be used to advantage for grading the trench ahead of the masons.

In bonding the new work with that previously laid (the

end of which should be toothed or racked back) all loose brick and mortar should be removed, and the brick cleaned and wetted before applying fresh mortar.

The arch of the sewer is built upon a "centre," which is removed when the arch is completed and the mortar suffi-

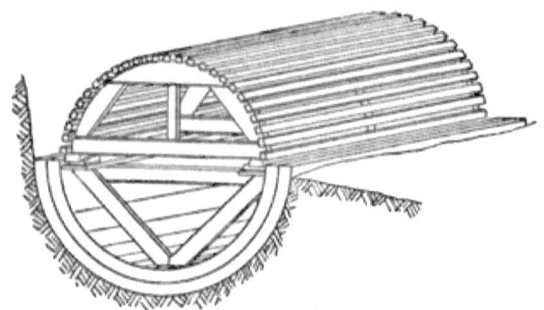

FIG. 25.—CENTRE FOR BRICK SEWERS.

ciently set. The centre usually consists of lagging supported by curved ribs of wood or iron. Probably the most common form is that shown in Fig. 25. To templets similar in form and general construction to those for the invert are nailed lagging-strips about 1 inch thick and 1½ inches wide, spaced 2¼ inches between centres, there being a templet at each end and intermediate ones spaced 3 or 4 feet apart. The lagging-pieces should be perfectly parallel, as their edges are used for lining the brick-work. If the radius of the arch exceeds 2 or 3 feet the lagging may be of 1½- or 2-inch by 3½-inch strips, spaced 4½ inches between centres; but the 2¼-inch spacing will give a better surface whatever the radius. The templets should lack 3 or 4 inches of being complete semicircles, so that when in position the bottom of the centre may be about 1½ or 2 inches above the springing-line of the arch. The centre may be held in position by a triangular frame under each templet, supporting a plank along each side of the sewer, upon which the centre rests, it being raised to exact position by wedges, as shown in Fig. 25. When the arch is completed the wedges are knocked out and the centre drops onto the two planks and can be pulled forward, sliding upon these. It is sometimes difficult, particularly with large and heavy centres, to draw them out, and to facilitate this a light temporary track has in some instances been built under the centre, which was placed upon wheels which rose 2 or 3 inches above the track when the centre was wedged up into position. By knocking out the wedges the centre drops onto the track and can be readily rolled forward. The use of rings of angle-iron to support the lagging gave good satisfaction in Denver, Col. (see Transactions of Am. Soc. Civil Engineers, vol. XXXV, page 113). For very large sewers it may be better to build each centre in place and take it apart in order to move it.

The arch should be built up at a uniform rate on both sides at once, and the last row of brick to be laid in each ring should be at the crown and should be driven tightly into place as a key. It may be necessary to split brick for this purpose, but it is better to have on hand a number of thin arch-brick (of wedge-shaped section), hard and tough, which will stand driving. The outside of the arch is usually plastered with $\frac{1}{4}$ to $\frac{1}{2}$ inch of mortar. The centre should be left under until the mortar is so set that there is no danger of the arch becoming deformed if it is drawn, the time varying with the character of cement, shape and thickness of the arch, and other details of construction. It is probably well in most cases to back-fill to the crown of the arch as soon as it is completed. But if the soil is wet, like muck, or if, when excavating-machinery is used, the buckets usually contain considerable water, no back-filling should be done until the mortar is thoroughly set.

If the arch is of stone or concrete masonry lined with brick the brick ring is laid as described above and the stone or concrete built on top of it. The arch is sometimes built of concrete without a lining, in which case the lagging-strips must be set close together. In the Wachusett (concrete) Aqueduct, 11 feet 6 inches in diameter, sheets of galvanized iron and zinc greased with black-oil were fastened over the lagging on the centres with good results.

After the removal of the centre the arch masonry will ordinarily be found somewhat uneven, with mortar adhering in flat lumps to a large part of its surface. These should be removed and the joints so pointed as to render the surface more even, or the whole inside of the arch may be plastered.

If there is masonry backing to the invert this is usually laid as uncoursed rubble or concrete up to within $4\frac{1}{4}$ inches of the invert-surface, the templet having been set to indicate this, and the brick lining is then laid as above described. If concrete is used and is not carried to the sides of the trench (see Plate VI, Fig. 9) a form of plank is used, inside which the concrete is rammed, and the plank removed when this is set. If the trench is sheathed and the concrete is built against the sheathing this cannot be pulled, but must be left in or cut off above the concrete. If stone masonry is used for invert-backing it is better to lay the course of stone next to the brick lining with radial beds.

If concrete is used for the entire sewer special forms must be made for each size of sewer, at least two sets being in use by each gang. The form for the invert may be made similar to an arch-centre, except that the lagging must make tight joints (its edges being bevelled to permit of this) and only the two or three on the bottom be fastened to the templets. This form is fixed in position, concrete is placed in the bottom, between the lagging and the earth, and rammed; one or two strips of lagging are then slipped into position on each side and concrete placed and rammed behind these; more strips are added and concrete rammed behind them, and so on until the concrete is brought to the springing-line of the arch. The forms should not be removed until the concrete is set. There is much danger that in this invert construction dirt and stones will get into the concrete, to its detriment, and great care must be taken to avoid this. The forms must be strongly braced down from the bank, to resist their tendency to rise when the concrete is rammed. The concrete should be just wet enough to permit water to be brought to the surface by light ramming. No heavy rammers should be used.

For making a concrete arch, if there is no brick lining, a centre is used with close lagging, or an open-lagged centre may be covered with sheet metal, as on the Wachusett

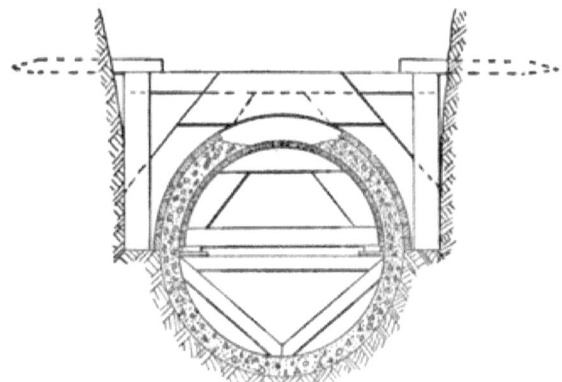

FIG. 26.—FORM FOR CONCRETE ARCH.

Aqueduct mentioned above. The outside form may be constructed as shown in Fig. 26, the forms being placed 3 to 5 feet apart, the lagging being loose and put in one strip at a time.

Concrete sewers have been built in a "travelling mould" (Ransome method), by use of which the entire sewer is constructed continuously, foot by foot. A core in the shape of a ribbon which can be readily withdrawn after use (Chenoweth system) has been used for small concrete sewers, to which the use of the ordinary centre and form is not adapted.

Concrete sewers are used extensively in Paris, Hamburg, Brussels, and many other European cities. Quite a number of American cities have built certain of their sewers of concrete, among these Washington, D. C., Reading, Pa., and Salt Lake City.

ART. 75. BUILDING MANHOLES AND OTHER APPURTENANCES.

These can most conveniently be, and usually are, built of brick. The foundation is sometimes of brick, but concrete is better in most cases. A stone slab set on concrete makes a good bottom for catch-basins.

The channel through a pipe-sewer manhole is sometimes built of brick, but a split pipe is better. If brick be used, the inside of the channel should be plastered with a coat of neat Portland cement. If any branch channel in a manhole is not to be used at once it should be temporarily closed to prevent deposits forming in it. The bench may be built up of brick plastered on top with cement, or of concrete. Or the whole manhole bottom may be of concrete, a wooden core being slipped into the opposite pipes and spanning the manhole to give the shape to the channel.

In leaving the manhole-opening in a brick sewer the end brick in every alternate course of the outside ring may be laid radially, thus presenting toothing protruding at right angles to the sewer-barrel. In this steps with horizontal treads can be built of brick trimmed to the necessary shape, from which the manhole can be carried up without danger of its sliding off the sewer.

To insure having the manhole of the proper size and shape a board templet may be used, being laid, in pipe-sewer manholes, upon the concrete foundation when this has set, and the brick started by it and carried vertically to the proper height. Another templet 24 inches diameter is fastened at the level of the top of the brick-work, its centre vertically above that of the bottom templet. Cords are strung from the edge of the top templet to the top of the vertical part of the brick wall, spaced about 2 feet apart around its circumference, and the brick laid to these. An experienced manhole mason, however, can build almost as symmetrical a manhole by eye only, and more quickly than if strings are used.

When the wall is about 2 or 3 feet high the benches and channels of the bottom may be constructed. It is well to lay plank in the bottom over the channels temporarily, to keep mortar and dirt out of them and out of the sewer during construction, as well as to hold the brick and mortar being used. The first step should be placed about 18 inches or 2 feet above the bench. When the wall is about 4 feet high four piles of brick, each 8 inches square and about 3 feet high, may be made on the bottom of the manhole and a platform of short loose plank be placed on these, entirely filling the manhole. This holds the mason, brick, and mortar until another 3 feet are built, when a second platform is similarly placed 3 feet higher. These are of course removed when the brick-work is completed.

The brick in a manhole may be laid as all headers, all stretchers, all on end with their edges exposed, or a combination of any two or all of these. Bats may be used in large or small proportion or not at all. A strong manhole can be built by using three courses of stretchers to one of headers, all whole brick, until a diameter of about 3 feet is reached, and from there to the top using three courses of squared bats to one of headers. The outside of the manhole should be plastered as the wall is built, since it may be impossible to reach it afterward. The head should be set as soon as the brick-work is completed, and the opening back-filled.

If the manhole is shallow, or for any other reason the diameter is to be rapidly reduced towards the top, this is ordinarily done by making each ring of brick a little smaller than the one below, the diameter of the manhole being reduced by 1 to 4 inches with each ring. Or it may be arched (Plate IX, Fig. 2), when the back-filling around it should be thoroughly tamped to assist in taking the thrust. In the case of flush-tanks particularly a flat iron ring is sometimes built in the outside of the brick-work at the bottom of the arch as a precaution.

Flush-tanks are built in a manner similar to the above. These, except at the very top, and catch-basin inlets, are usually larger in diameter than manholes, and are built throughout of whole brick. Extra care should be taken to have all joints filled with cement and tight, and the work well bonded. After the cement in flush-tanks and catch-basins has fully set they should be given on the inside two or three washes of neat-cement grout, laid on with a whitewash or similar brush, care being taken to cover the entire surface with each coat, which should be allowed to dry before the next is applied. This will seldom fail to give a tight wall.

No water should be turned into the trench for flushing or other purposes before the cement in these appurtenances, as well as in the sewer, has set.

If masonry in either sewers or their appurtenances is laid in freezing weather special measures and precautions should be taken. The sand, stone, brick, and water should all be heated before being used, and special care taken to see that no ice or frozen dirt is in the mortar, on the stone or brick, around the sub-drain, under the pipe, or under or behind the brick or concrete sewer-invert. To insure the last it is well to take out the last foot or two of trench just before the sewer is to be laid in it. If any frozen earth is found under the sewer grade it should be removed and replaced by sand or gravel thoroughly rammed.

The water for mortar can be conveniently heated by injecting into it steam (as the exhaust from a pump- or excavator-engine), it being kept in several hogsheads or oil-barrels. The brick and stone can be heated by piling them as in brick-kilns and burning a wood fire under them; and the sand by being piled over these, or in large iron pans such as are used for heating asphalt.

ART. 76. FOUNDATIONS.

Piles are ordinarily used for sewer-foundations in soft soil. They usually support a timber platform, but in some instances concrete is placed directly upon and around their heads. For driving them the ordinary pile-drivers are used, or they are sunk by the water-jet. If they are to support platform timbers they must be driven carefully to line and sawed off accurately to grade. It will sometimes be advisable to drive the piles before the excavation has proceeded very far, using piles considerably longer than actually required, as the jarring of the banks of the trench may thus be avoided, as well as the inconvenience of moving the driver through or over a trench full of braces. The objection to this plan, aside from the cost of the additional length of the piles, is that they interfere with the excavation.

In moving an ordinary pile-driver through the trench it will be necessary to remove the braces ahead of it. But no brace should be removed until another has been inserted behind the driver-frame between the same rangers and as close to the first as possible. This trouble might be avoided in many cases by placing the pile-driver on a track, on a level with the ground, over the centre of the trench; or the track may be on the surface at one side of the trench. The driver is then provided with movable hammer-guides, which can be lowered into the trench and raised with ease. The use of the steam-hammer pile-driver is often advantageous, and in sandy soils the water-jet can be used to advantage. Neither of these last is interfered with in its operation by the bracing.

The dimensions and construction of the platform follow the rules for ordinary foundations. There is usually but one set of timbers under the planking, which is in most cases composed of one or two layers of 2-inch to 4-inch plank, as in Plate VI, Figs. 3, 5, and 6; although in some instances heavy timbers are used, as in Plate VII, Fig. 10. Any timber which is to be placed where it will not be continually wet should be creosoted.

If a platform is used without piling, sills, longitudinal or cross, should be placed under the planking, although in the case of small sewers these may consist of lengths of 2-inch plank only. Platforms without piling or heavy sills are of little permanent service under large sewers, but during construction may serve to prevent local distortion of the masonry before the cement has set. One or two lines of plank placed lengthwise under a pipe sewer, however, are in many cases of permanent value, back-filling being thoroughly filled and rammed between the pipe and the plank.

Among the best of our woods for foundations are the cedar, oak, elm, alder, and beech. All bark should be removed and the sap dried out from piling or sawed timber. The platform timbers should be fastened to the piles with iron drift-bolts or treenails.

ART. 77. PUMPING AND DRAINING.

Next to quicksand, water is probably the worst enemy of the sewer-contractor and requires a large share of the attention of the engineer. If there is but a small trickle or ooze of water into the trench it may interfere but little with the excavating, and will collect at points in the bottomed trench whence it can be removed at intervals by a bucket. If the amount becomes somewhat greater it may still be handled without the use of sub-drains, that from where the pipe has been laid being shut off by the back-filling.

The amount from the trench ahead of the sewer may need to be pumped, however. For removing small quantities of water from a trench probably nothing is better than a diaphragm-pump. Tin "boat-pumps" are often used, but will not handle so much water, are less economical of power, and are not so convenient as the diaphragm-pump; they can, however, be used in trenches more than 20 feet deep, where the diaphragm is hardly practicable. Under favorable conditions a diaphragm-pump can be made to raise 5000 or 6000 gallons per hour. Diaphragm-pumps can be used in deep trenches by placing a second pump upon a platform half-way down the trench, which discharges the water into a tub, from which the first pump raises it to the surface. Or the upper pump may not be used, but a trough may carry the discharge from the lower one to an opening in the sewer at a point where the cement is so set as to be uninjured thereby, the water flowing through the sewer to its outlet.

A sump-hole of ample size should be made in the bottom of the trench to receive the suction-pipe, which should be provided with a strainer at the bottom. If the material is sand or soft ground it is well to place a pail or keg in the sump to keep the end of the suction-pipe from being buried, the top of the pail being just below the level of the trench bottom. The pail should be watched and material kept from running over its edge. The excavation should usually be so carried on that the whole trench slopes toward the sump-hole, each laborer seeing that the water flows through his section to the next lower.

Where a sub-drain is being laid the water is frequently permitted to flow from the trench under excavation to and through this. In many if not most soils this is bad policy, since it leads to a silting up of the drain by the large amount of material washed in from the trench. It is better in most cases to leave or make a dam at the upper end of the completed trench, and place a sump-hole just ahead of this and below grade, from which the water is pumped. When a section of 20 or 30 feet has been excavated to grade another dam and sump-hole can be placed at the head of this section and the others removed, the sump-hole being filled with sand or gravel or other good material well rammed.

Where a sub-drain is started from a sump-hole, or that lower down the line is found to be too small to carry the water coming to it, a pump must be placed at this point also to remove the water from the sub-drain which is to be laid beyond it. This water is frequently raised to the sewer only, the pump being placed in a manhole and discharging the water below a temporary dam in the sewer, which prevents its flowing up the sewer onto the work.

Two or more hand-pumps are sometimes concentrated at one point when the amount of water is considerable. It would in many instances be cheaper to use a steam-pump at such a place. Piston, centrifugal, and wrecking pumps, pulsometers, and steam-siphons are the steam appliances in most common use on sewer construction. In all of these iron suction-pipes are used, from 4 to 8 or 10 inches in diameter. The piston-pump is the most economical, and adapted to widely and rapidly varying quantities of water, and if the water is fairly clean needs very little attention. It cannot, however, pump gritty water without rapid deterioration. The centrifugal pump can raise muddy or gritty water, chips, and even small stones, its first cost is less than that of a piston-pump, and it can be repaired more cheaply if damaged. It requires a fairly constant and fixed quantity of water to keep it working, and is apt, especially when a little worn, to give trouble by losing its priming, when the rising of water in the trench before it can again be primed may give trouble. The wrecking-pump the author has found to be an excellent pump for sewerage-work. It will lift and discharge anything which can pass through its suction-pipe and is extremely simple in action. All these pumps must be firmly set over or near the trench and their position can be changed only with considerable labor. It is better to set them directly over the sump and have a suction-pipe as short and with as few joints as possible.

The pulsometer pumps muddy and gritty water, but is not economical of steam and, except in experienced hands, is apt to act in a provokingly contrary manner, particularly after some use. It has the great advantage, however, of portability, being suspended by a chain, which permits rapid changing of its position without cessation of pumping, the steam being conveyed to it through a rubber steam-hose. For pumping large quantities of water at the point where excavation is proceeding and where frequent change of location of pump and suction is necessary it is perhaps the best contrivance on the market. The steam-siphon is likewise conveniently portable, but is most extravagant of steam and is hardly practicable for raising large quantities of water.

The pulsometer and siphon are particularly adapted to raising water from the point where the work is progressing with the least interference therewith. Piston, centrifugal, and wrecking pumps are best used at a distance from the work to lift water which has flowed to them through sub-drains or the sewer, although they are often used at the work when the same sump can be used for two or three days at a time.

All suction-pipe on either steam- or hand-pumps should be provided with a strainer at the bottom, and the centrifugal requires a foot-valve, which it is also well to supply for the other steam-pumps. If a chip or other obstacle should hold this valve open and prevent priming the suction a shovelful of stable manure dropped into the suction-pipe will in many cases enable the valve to hold its priming.

All parts of the machinery should be readily accessible, particularly any valves, and wrenches and screw-drivers, packing, oil, waste, duplicate nuts, washers, etc., should be kept constantly at hand. A cessation of pumping for 15 minutes may permit the water to drive the workmen from the trench, to soften the banks and endanger the sheathing, ruin the green masonry, stop up sub-drains, or do other serious damage. A good, intelligent, careful stationary engineer is a necessity on such work.

The water raised from the trench should not be discharged upon the ground near the sewer, unless the street has impervious pavement, as it might soak back into the trench and be pumped over and over again. It may be carried to the nearest watercourse or sewer-inlet or manhole along the gutters, in wooden troughs, or in sewer-pipe temporarily laid on the ground with joints tightly calked with oakum or clay.

It usually pays to keep the water pumped down all night, even if there is no work to be damaged by its rising, as this would again fill the surrounding ground with water, which might not drain out for several hours after pumping began the next day. It may be well to whitewash one or two sheathing-plank down to the trench bottom each evening, which will give evidence next day if the engineer has not kept the water down. A shelter should be built in front of the boiler to protect the engineer from storms.

While using a diaphragm-pump always have spare diaphragms and an extra length of suction-hose on hand.

Moving a pump and boiler often costs more indirectly in interference with the work than the immediate expense comes to. In general every effort should be made to set the pump in such a place and manner that it need not soon be moved. Be sure to have the blocking under it solid, to prevent the suction-pipe joints from working loose or breaking.

ART. 78. HANDLING WET AND QUICKSAND TRENCHES.

If excavation is in good material and of comparatively uniform depth a sewer gang once organized should move along at a uniform rate of 300 or 400 feet a day for small pipe sewers, 25 to 200 feet for brick ones, and with little but routine work for the foreman. If genuine quicksand is encountered, however, every foot of progress must be fought for with unflagging energy, pluck, and intelligence. In ordinary wet trenches the difficulty, while not usually so great, is sometimes considerable. In both an intelligent adapting of the work to every new exigency is necessary.

Water is met with as springs in the trench or as a general exuding from all the ground. The former can easily be managed by catching the water at its point of exit and pumping it away. If it enters from the bottom of the trench it can sometimes be caught in a trough and led back and discharged into either the completed sewer or into a tub in which the suction-pipe of a pump is placed. It is absolutely useless to attempt to stop the water from coming out of the ground; the endeavor must be to handle it after it gets out. In the case of a spring in a brick-sewer trench a method often advantageous is to build into the brick-work opposite the spring a small pipe, 2 to 4 inches diameter, through which the water can enter the sewer, and to conduct it back from there to the finished sewer in a trough. This pipe can be plugged after the masonry is thoroughly set, but might better be left open to drain the ground if in a storm-sewer, or if in a combined sewer and well above the line of flow of house-sewage. This pipe can, in many cases, be so driven into the bank at the spring that the water will flow through it and the trough be set before the brick-work is begun at that point, the trench being thus left dry.

If the water does not enter as a spring and consequently cannot be caught in this way, but if the ground is a gravel or is not readily softened by the water, an outer ring of brick may be built with quick-setting cement, and plenty of it in beds as well as joints, an occasional brick being left out to permit the water to enter the sewer-invert, over which it can flow to a sump-hole ahead or through the sewer below. If openings are not thus left in the brick-work the water will force its way through the joints. Plank should be placed over the brick-work as fast as it is laid for the masons to stand upon. This outer ring when set may be found uneven of surface, but the joints will probably be tight. The openings may then be closed by inserting a brick and calking the joints with cloth, oakum and cement, wooden wedges, tea-lead, etc., or a pipe may be inserted and the water allowed to enter it as described above. The outer ring being thus made water-tight, the inner ones can be built as usual, any depression in the outer ring being well filled with mortar. In this and in all brick-, stone-, and particularly concrete-work which water flows over while green the surface can be protected from wash by spreading rather heavy, strong brown wrapping-paper over it. Cheap wood-pulp paper is of little use. The paper when wet will cling to the masonry, remaining intact for days and even weeks.

Another plan is to dig a sump-hole $1\frac{1}{2}$ to 3 feet deep in the centre of the section of invert under construction, and keep the water lowered in this by a pump until the brick-work is completed and set everywhere except over the sump. If the ground is very porous the water will all flow to the sump and leave the trench dry for several feet in each direction.

When the surrounding masonry has set the suction-pipe is removed from the sump-hole, and this is filled with sand, gravel, or concrete, thoroughly rammed. The remaining brick-work is then laid, with or without a pipe through it, as described above.

A better plan is to use sub-drain pipe, discharging into a sump, which is to be pumped if there is no outlet for it or if the drain below is too small to carry all the water. A disadvantage in this connection of building either brick or pipe sewers down instead of up grade is that the water cannot be run away through the sewer or sub-drain, whether it be pumped or not, and although it drains away from the work it is only to soak into the ground ahead and make that all the wetter, besides the fact that it is accumulated where the excavation is in progress. Not only this, but the ditch acts as a drain to conduct down to the work water from all the territory above which has been passed through, the use of a sub-drain adding greatly to this amount. If the trench be dug up hill it will while advancing tend to drain out the ground ahead and a trench may be found dry which would be wet if approached from above. In some instances where a trench has been extended up to ground which seemed hopelessly wet, and the trench thoroughly braced and left open for a week or two, the excavation was then resumed without difficulty, the ground being found comparatively dry.

This fact, that wet ground will in many cases drain out if an outlet be provided, may be taken advantage of in several ways. For instance, if beneath the wet porous soil, but above the sewer grade, is a stratum of clay the trench may be carried down to this, braced, and allowed to drain out, when the clay can be readily cut out dry instead of as a thick, sticky paste which mires the feet and will not leave the shovel. Quicksand can sometimes be dried out if the water be given an outlet and sufficient time allowed. It will then be almost as hard as rock, but much easier to handle than in its quick state.

If sewer construction is in the shape of an extension from a line already in use into which the water must not be run, or if it is carried on in sections which have no outlet, a pumping-station can take the place of an outlet, or a ditch can sometimes be carried to a watercourse lying below the sewer. The latter is always the better plan if not too expensive, as there is then no danger from broken or disordered pumps. But the ditch must be above the reach of any possible flood in the stream into which it discharges.

A plan used with success on the Metropolitan Sewerage System (Boston) and elsewhere is to drive 2- or $2\frac{1}{2}$-inch pipes by water-jet on one or both sides of the trench, 10 to 15 feet apart and to a point 2 or 3 feet below the bottom of the sewer, and, by connecting a number of them to a 6-inch suction-main and pumping on them for a few days, lower the ground-water before the excavation reaches this point, and keep it lowered until the work here is completed. If the trench is less than about 20 feet deep the pipes may be driven outside the trench, but if more it will probably be necessary to put them and the pump inside the sheathing at a distance of not more than 20 feet from the bottom, although they may be in the way there. The sinking of such tubes in Newton, Mass., cost from 8 to 50 cents per foot.

In laying pipe sewers in wet trenches much of the above is not applicable. The best method for such work is the use of sub-drains. When the ground is not excessively wet the trench is then dry for the laying of the sewer-pipe. But where there is a large flow of underground water it may be impossible for it to reach the sub-drain, through the overlying gravel or stone, as rapidly as it enters the trench. Frequent sumps must then be provided, with a pump at each, there being always one only a few feet ahead of the sewer. If water still flows over the trench bottom to the sump it may be necessary to lay the sewer in concrete. In fact this is always desirable, though expensive, in wet trenches or where sub-drains are used. In using concrete it should be placed and rammed in the trench and the pipe bedded in it before it sets. The concrete may be brought up only a short distance above the invert of the pipe, being sloped down toward the sheathing and forming a gutter on each side in which the water may run to the nearest manhole or sump. If this flow is considerable plank or boards or heavy paper may be laid on the concrete to protect it from wash. The rest of the sewer-joint may be made in the ordinary way. It

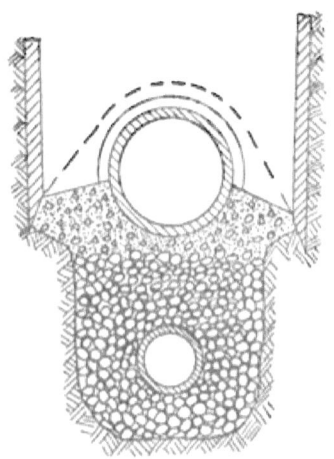

FIG. 27.—SEWER-PIPE LAID IN CONCRETE.

is better, however, to also carry the concrete entirely over the sewer at the joints after a stretch between manholes is completed and the side gutters are no longer needed.

Water should never be allowed to stand in bell-holes after a pipe is cemented. If liable to, the bell-hole should be filled with cement or concrete, or at least with sand or gravel well tamped. No water should be allowed to run through a sewer until the cement is fully set. Particular attention should be paid to branches and slants in wet trenches to see that they are tightly sealed. It is an excellent plan to build a dam at each end of a stretch of sewer in a wet trench, after the sewer is completed and cement set, and before back-filling above the pipe, and allow the water to stand upon it. Leaks thus discovered are then readily accessible for repairs.

In moderately wet ground it is often advisable to place

dams across the trench at intervals of 15 to 30 feet, that there may not be so great a stream continually flowing by the men while working. The head of the trench being kept on an incline, water collects above each dam until there are no dry places left in the sections in which to dig, when the dams are opened in succession, beginning with the lowest, and the water flows to the sump, from which it is pumped. The dams are then closed and digging resumed immediately above each, the laborers moving up the slope as the water rises above each dam.

The combination of water with a particular kind of sand produces what is called quicksand. Any object resting upon this sinks slowly into it until it has displaced its own weight of sand. But a pick can hardly be driven into quicksand which has not been disturbed. The sand is very fine and is easily stirred up and carried by running water, but will quickly settle into a tough, compact mass which, if allowed to dry out, will become almost as hard as soft sandstone. Quicksand is semi-fluid and will run under sheathing unless it be driven to a considerable distance below the bottom of the trench. If the influx is not cut off by deep sheathing, by the time the excavation is 2 or 3 feet into quicksand a point is reached beyond which no headway can be made, the bottom remaining at the same level however much be taken out of it. After a time the cavities behind the sheathing, caused by the flowing of the quicksand from there into the trench, permit the ground-surface to settle or to drop entirely, and the sheathing, relieved of outside pressure and friction, is apt to completely collapse. If there is any possibility that such a cavity is forming all braces should be nailed to the rangers and tied together by cross-bracing, and outside rangers braced against the sheathing from the curb or other points well back of the trench. If there is more than one course of sheathing the plank in the upper ones should be nailed to the rangers.

If the ground-surface should fall into the cavity thus made sod, straw, brush, etc., should be thrown against the sheathing, which will stop the quicksand from flowing into the trench. The entire cavity should then be filled with earth, ashes, or some good filling material. It is in most instances well, if the condition is such as is shown in Fig. 28, to remove

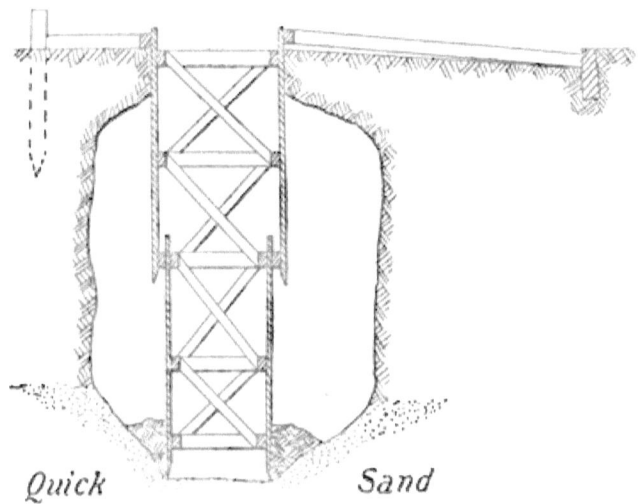

FIG. 28.—SHEATHING A BADLY CAVED TRENCH.

a plank or two here and there from the upper course of sheathing and throw into the cavity sods or straw, and then, after bracing the sheathing as above described, to break down the top soil and fill the cavity with good earth. Fig. 28 is no exaggeration of conditions sometimes occurring in quicksand. A preventative, which is usually effective, is to keep several men continually driving the sheathing with light mauls, or better still keep steam-hammer pile-drivers at work, so that the bottom of the sheathing is continually maintained a foot or two below the bottom of the trench. This is a precaution which should never be neglected.

One effect of the formation of the cavities described is that the top earth tends more than ever to fall towards the trench, and consequently the strain on rangers and braces becomes severe. It is better to multiply the number of rangers than that of the braces to each ranger, as the trench is then less obstructed for lowering materials.

Quicksand has usually only a little water flowing through it, but that little should be handled by pump if possible and not allowed to run into the sewer or sub-drain, the result of which would be the rapid choking of the drain, or of the sewer if small, by quicksand. Quicksand should not be thrown directly back upon the finished sewer, as its angle of stability is exceedingly small, and it is apt to run forward to the upper end of the sewer, either requiring to be handled over again or flowing into the mouth of the pipe. If thrown upon the bank and dried out, however, it becomes very hard and expensive to shovel back. It is probably better to back-fill with it immediately at some distance from the sewer under construction, carrying it there by excavating-machinery or in wheelbarrows, or to let it partly dry upon the bank before throwing it back. It is well to have a few short plank nailed together to form platforms which can be placed in the trench bottom for the men to stand upon, as otherwise they will lose much time digging themselves and each other out. The length of open trench should be kept short and the men worked as close together as practicable, and sub-drain with its gravel and platform put in and sewer laid as rapidly as possible. Even then the danger is not over, as the structure is liable to be raised out of place by inflowing quicksand. A pipe sewer which had been laid in quicksand and covered with the same material as back-filling has been known to rise more than 3 feet overnight, practically floating to the surface of the quicksand. To prevent this a plank may be laid over the sewer and braced down from the sheathing and the pipe thus held in place.

In the case of a brick sewer the platform should generally

be set upon piles (which can best be driven by the water-jet) and immediately loaded with brick or stone which is to be used in the construction. It is advisable in most cases of large sewers built in quicksand to place close sheathing across the trench, 15 to 30 feet ahead of the completed sewer, making a coffer-dam, inside of which the next section of sewer is built. Meantime other cross-sheathing, 15 to 30 feet still further ahead of the last, is being driven, together with the side sheathing, and the coffer-dam thus formed excavated. When this is down to grade and the foundation in, the cross-sheathing just ahead of the completed sewer is removed and the sewer continued into the next section of trench. In each of these coffer-dams, usually in one corner, is a sump from which the water is kept pumped. A pail or barrel should be used in every sump in quicksand and the sand kept dug away from around it, as it is very apt to reach and stop the suction-pipe, from which it is difficult to remove it. Gravel or fine broken stone may be placed around the barrel and carried a few inches above it to prevent the sand reaching it.

Laying pipe sewers in quicksand may be even more difficult than building brick ones. If a platform foundation is used there is not sufficient weight in the pipe to hold it down and it must be strongly built and braced down from the sheathing. The following plan has worked well: A sill of 4×6 timber is laid near each side of the trench, which has been brought as near as possible to grade, and a short piece of plank is stood upon it near one end. Two or more men then stand upon the sill near this end and work it down to the necessary depth, when the upright is nailed to a brace or ranger and the sill at this point thus held down to place. The other end is then worked to place and similarly braced and the other sill treated likewise. By this time the sand is probably several inches deep over both sills. Cross-planking for the platform having been sawed to length—the closer they fit between the sheathing the better—one at a time a place is cleaned for them and they are nailed to the sills with close joints. Good material is then placed on this platform and the pipe laid thereon and the same material immediately backfilled around and above it. When the pipe has been laid to the end of the platform it should be tightly plugged, if the next platform is not ready to continue laying (as it probably will not be), to keep out the quicksand, which may rise above the pipe-invert before the laying is continued.

Another plan for laying sewer-pipe, and an excellent one for sub-drains, in quicksand is as follows: Two planks, each about 6 feet long, are stood upon edge a sufficient distance apart to permit laying between them the sewer-pipe, or drain-pipe and required broken stone, and a strip of wood is nailed across their tops at each end. The other edges are then turned up and similarly treated, a bottomless trough being thus formed. A loose bottom is provided to fit it, usually in two or three pieces. The bottomless trough is then placed in the trench with the plank on edge and worked down into the quicksand until to the necessary depth and in the correct line, and is braced down from the sheathing. The sand is then shovelled out of this and the bottom planks put in, one at a time, the men standing on the trough bottom until all the planks are in and secured, which is effected by placing a cleat across the bottom at each end and fastening it to the sides of the trough. The sewer, or sub-drain and broken stone, are laid in this and good material is packed around and over the sewer. This method is also adapted to dry running sand, where the trough can generally be used without a bottom.

Another method sometimes employed is to excavate the quicksand in short sections somewhat below the pipe-level and refill it with cinders, or spread burlap over the bottom and cover it with gravel, on which the sub-drain or sewer is laid.

Still another plan is to drive tubes a few feet apart throughout the entire trench, a few at a time, their bottoms being all about a foot or two below the pipe grade, and inject cement and water under pressure, the cement filling the interstices of the surrounding sand and forming an artificial stone, which prevents the quicksand from rising. The pipes are then removed and the trench excavated. This process is patented and expensive.

In laying sub-drains through wet soils, particularly wet sand, a great deal of annoyance and expense will almost surely be incurred if some plan is not carried out for keeping the pipe free from deposits, which will form from the dirty water flowing into the end of the pipe. Probably the best plan is to keep a rope drawn through at least 600 feet of the pipe, the end being drawn forward through each length as it is laid. At intervals of from 10 minutes to half a day the rope should be pulled back and forth to stir up the deposit and keep it moving. It is well to knot up a light chain and at least once a day tie it firmly to the rope and draw this through the drain a few times. If the rope is neglected for too long a time it may become imbedded in the deposit and require six or eight men or even a team to draw it through the pipe. It should be amply strong for this. When a section between manholes is completed the rope should be left in until the next section is completed, as a part of the dirt flowing through the upper will probably settle in the lower section. The latter should be cleaned at least once every day. It may be well, if the deposits are considerable, to fasten to the rope in the lower section a strap to which two or three tapering tin cans are riveted (see Fig. 29). These are drawn a little way into the pipe, closed end first, and then drawn back and the dirt emptied from them. (If started through the pipe mouth first they would probably pile the dirt ahead of themselves into an immovable mass.) When not in use these should be kept out of the pipe, as should also the knotted chain above mentioned, to permit the free passage of water.

It is advisable to flush the drain with comparatively clean water as often as possible. This may be done by catching the ground-water by dams, as described above, and, when the drain has been laid up to a dam, bailing the water rapidly from this into the drain. Or a plank may be set across the trench bottom just ahead of the pipe, so as to catch any mud or stones which may be washed down but permit the water to

FIG. 29.—APPLIANCE FOR CLEANING SUB-DRAINS.

flow over, and the dam be broken, but kept under control, so that it can be closed if desired.

It is well to always keep in the end of the pipe a pan with fine holes over the upper half of its bottom, these holes forming the only entrance for water to the pipe. Sticks and stones are thus kept out of the sewer, as well as the water most thick with mud. If the water is clear the perforated part of the pan can be placed at the bottom. These remarks refer particularly to sub-drains, since water should not be permitted to rise above the sewer-invert.

The entire system of sub-drains cannot be flushed too often during construction—by hose from the fire-hydrants, if possible. If a drain is stopped up in which there is no rope, or the rope cannot be moved, a hole can sometimes be forced through the obstruction by a line of $\frac{3}{4}$-inch or 1-inch iron pipe. If this is fastened by a special bushing or otherwise in the end of a hose and water forced through it, it can be driven through in almost every instance if the friction between the iron pipe and the deposited material does not become too great. If the section in which the stoppage occurs is not at the incompleted end this plan cannot be adopted; but an old length of 2½-inch hose with no coupling on one end can be placed at the end of a line leading from a fire-hydrant to and down a manhole, and this end pushed into the drain; when the water is turned on the hose can be pushed forward as if it were a flexible rod, and the water from the hose will wash the obstruction loose and bring it back to the manhole, where it can be removed from the sub-drain well by hand, the water rising up and overflowing into the sewer. Sand, gravel, and even brick-bats have been washed out of drains by this hydraulic process.

When laying a pipe sewer the manhole is not usually constructed until the sewer has been laid on each side of it. In quicksand if the trench is opened through where the manhole is to be it will immediately fill up above the sewer. The pipe must therefore be plugged at the end. It is, for this and other reasons, often desirable in quicksand to build the manhole before the sewer reaches it, openings for the sewer being left in the manhole-wall at the proper points. The excavation for the manhole must be made in a well-hole close-sheathed for at least a foot lower than the sewer-invert. It will be found difficult to get the bottom in with the ordinary methods, particularly if there is a sub-drain well to be put in. In such a case the following plan has been used with success: A 12- or 15-inch pipe with two T branches of the size of the sub-drain, temporarily plugged, is lowered into position to act as the sub-drain well, the bell being up and the branches being placed at the grade of the sub-drain, which connects into them. The manhole excavation having first been carried to the depth necessary for the foundation, this pipe may be lowered by resting upon it with the knees and digging the sand from the inside, care being taken to keep it vertical and in the proper position. When it has reached the required depth the sand is scooped out a little below its lower end and one or two bucketfuls of concrete placed there and rammed (see Plate X, Fig. 9). It is well to place a board bottom inside the pipe on top of the concrete and to place brick on this to keep the concrete from being forced up, the brick being removed after the concrete sets. If necessary another 12- or 15-inch pipe is placed upright in the hub of this one. A length or two of drain-pipe is fixed in each branch of the sub-drain well in a horizontal position and in the proper line to connect with the sub-drain when laid, and the manhole bottom is then dug out to the grade of the bottom of the foundation and concrete placed there, before the sand rises, in small areas of 8 or 10 feet at a time. This is done rapidly and the concrete loaded with brick, if necessary, to hold it down. The concrete is placed last where the channel comes and a split-pipe invert is at once forced down in it to the proper grade, and a straight-edged plank placed on edge in the invert bottom and braced down from the sheathing to hold it in position. The formation of the manhole bottom is then completed and the walls built in the usual way, sewer-pipe being built into the manhole-walls where the sewers are to enter it, but loosely enough to permit of sliding the pipe out. The sewer already laid or to be laid is carried through this opening by a pipe cut to the necessary length, the sheathing having been cut away here to permit this.

Another plan is to lay a plank foundation for the concrete, one plank at a time being put in place and fastened to the sheathing, thus forming of the whole a tight box, in which the manhole is built. Flush-tanks, inlets, and other appurtenances can of course be built in the same way.

ART. 79. RIVER-CROSSINGS AND OUTLETS.

For convenience of inspection and as permitting easier maintenance it is best to carry a sewer across a stream on a bridge or trestle, keeping its invert at the hydraulic gradient;

unless, of course, this is below the river-bed, when the sewer will occupy that position. Very often the use of bridge or trestle is impossible or prohibitively expensive, and then an inverted siphon is necessary. In either case the pipe will probably be of iron or wood, although a combination of these with masonry is sometimes used. In some instances it may be better to build the siphon in tunnel, when it should be lined with brick or concrete; or, as is usually better, two or more iron or wood siphon pipes may be laid in the tunnel, easy access to them being thus afforded.

A bridge or trestle for supporting a sewer should seldom be built of wood, owing to the difficulty of providing for the sewage when necessary renewals are being made. It may in some instances be unsafe to support a sewer by an existing

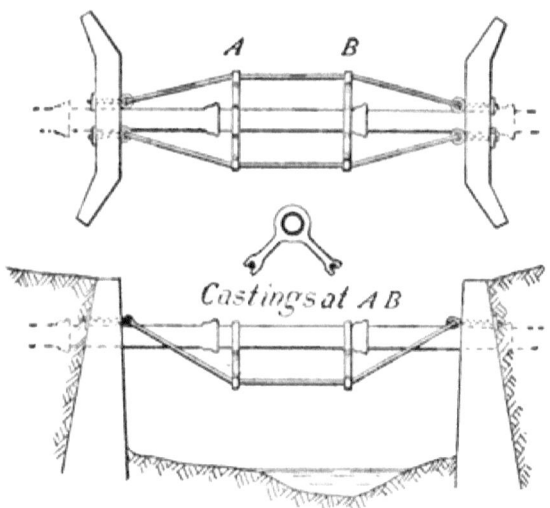

FIG. 30.—SEWER CROSSING CREEK ABOVE WATER.

bridge, owing to the great increase of load thus brought upon it. (An 18-inch cast-iron pipe flowing full of sewage will weigh about 225 pounds per lineal foot.) The bridge has in some cases been relieved of this weight by constructing the pipe in the form of an arch, but this is not generally advisable. A simple design for a short span, as over a creek, is shown in Fig. 30; or the pipe could be supported inside an iron or steel box girder of suitable size and strength. There is no danger of the sewage freezing unless the pipe is exposed for a stretch of many hundred feet.

If the distance to be crossed is more than 200 or 300 feet the inverted siphon will in most cases be found advisable. Its construction under water will be similar to that of river-crossings laid to grade, except that in the latter the most advantageous depth cannot be chosen.

The joints and pipe of subaqueous siphons and other sewers should be perfectly water-tight, as it will be necessary at times to empty them of sewage for inspection. They should, if of small pipe, be laid to as straight a line and grade as any part of the system. If they are sufficiently large to be entered this is not so important. They should never be laid on the bed of the river, but always beneath it.

For a sewer up to 30 or 36 inches diameter cast-iron pipe with lead or hardwood joints may be used. The trench is excavated at least 18 to 24 inches wider than the pipe and 6 to 12 inches below its grade. Inside this trench the pipe is placed and suspended to grade or blocked up at intervals. The joints are made and concrete is placed under the pipe at all points, completely filling the trench for a distance of 2 or 3 feet above the pipe, or to the surface of the river-bed, it all being thoroughly rammed. If the concrete does not reach the bed of the river it is well to throw loose stone over and along each side of it.

If the river is not very deep, or a time can be chosen when such is the case, it is in many instances practicable to confine it to half the width of the bed, at the point of crossing, by an earthen embankment or timber coffer-dam, or combination of both, carried, just up stream from the line of sewer, from above the water-line out to mid-channel, across the line of sewer, and back again to the bank a few feet lower down. The enclosed space is then pumped out, the trench dug and sheathed, the pipe and concrete put in position and covered, and the dam removed and a similar one placed upon the opposite side of the river, which then flows over the pipe already laid. In many cases the best form of dam for sewer-crossings is made by permitting the close sheathing of the pipe-trench to serve also as a dam, extending above the water-surface and backed by earth embankment. A brief statement of the details of carrying out this plan, which must, however, be varied under different conditions, is given.

The sewer having been laid up to the river-bank, a stout stake is driven into the river-bed about 10 feet from the end of this and in line with the down-stream side of the trench. If necessary another is driven a few feet lower down and a brace set from the foot of this to the top of the former. A frame of rangers and braces is built upon the bank, of dimensions proportioned for the proposed trench, and floated to place in line with the trench already dug, the inner end being fastened in position against the end of this trench and the outer being held by the stake just mentioned. Sheathing is then driven on both sides and the end of this frame (the end braces are flush with the ends of the rangers) as deep as is possible before excavating is begun, and earth banked against the outside of it. The water is then pumped out and the trench excavated, the sheathing being kept driven as low as possible, additional rangers and braces being added, and the excavated material thrown just outside of it. When this trench is at grade the pipe is laid, concrete put in, and trench back-filled ahead to cross-sheathing which has been set just back of the end of the pipe. Another frame has meantime been started just ahead, sheathing driven, and outer embankment made. The cross-sheathing between the new and the completed trench is drawn and the excavation continued.

Cross-sheathing is set at frequent intervals and the trench filled up to it to reduce the length of open trench which must be kept free of water. It is advisable not to cut off the sheathing and remove the embankment at any point until the construction is completed to mid-stream. It will usually be necessary to keep a pump going constantly during construction. If the stream is subject to freshets it may be well to set the pump upon a flat-boat anchored against the up-stream sheathing. The boiler may be kept upon this boat or upon the bank, the steam-pipe in the latter case being carried along the sheathing.

If the bed of the river is gravelly considerable trouble may be experienced from water leaking into the trench, the entering water having perhaps passed into the ground many feet from the sheathing. The embankment may in such a case be carried as far as possible from the sheathing on every side, or a thin layer of fine sand, sandy loam, or loamy clay may be spread over the bottom for 50 or 75 feet above and below the trench. Also manure, brewery-meal, etc., has been used to stop up the pores of the gravel. Heavy, closely woven canvas is excellent for use in such a case, in large squares or strips tightly sewed together, one end being fastened above water against the outside of the sheathing, the other anchored by stones or other weights beyond the part of the bed which is giving trouble.

An excellent material for the embankments is a puddle of clay, sand, and gravel. Clay alone is almost useless. Fine and coarse sand mixed, with or without gravel, is better than clay alone. All sticks, roots, and large stones should be removed from this puddle, and anything which, reaching through the embankment, may offer a course for the water. If puddling material is scarce a double row of sheet-piling may be carried around the work, the two rows being from 2 to 5 feet apart, braced together only at the top, and the space between them filled with puddle well worked and rammed. Experience in this class of work is almost essential to its proper prosecution, and written directions can give only the barest outlines for meeting but a few of the difficulties which may be encountered. Pluck, foresight, a fertility in expedients, and common sense are prime requisites for this work. The water must never for an instant be allowed to get the upper hand. If nothing else is at hand the very clothes off one's back should be taken to stop a leak temporarily, should

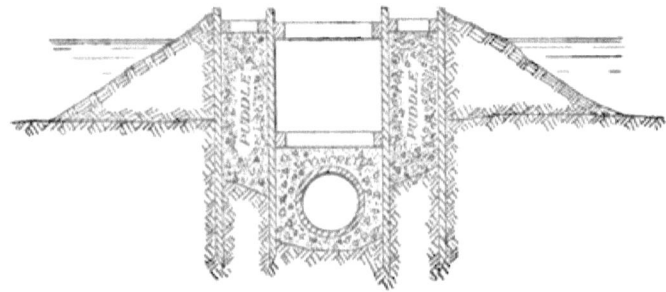

FIG. 31.—COFFER-DAM PUDDLE-WALLS.

one unexpectedly develop in an embankment. Never permit a brace, stick, or any object to extend through an embankment or puddle-trench or -wall. If a trench surrounded by water shows signs of collapsing from outside pressure and no material for additional rangers and braces is at hand the trench can sometimes be saved by allowing water to fill it, and then, when the material has been obtained, the water can be pumped down and bracing put in as it lowers. But this is a somewhat desperate remedy.

An outlet for the Massachusetts Metropolitan Sewerage System at Deer Island, in 5 to 10 feet of water, was built in open trench, with double sheathing and puddle, as in Fig. 31. The sewer was 6 feet 3½ inches inside diameter and the trench 10 feet wide on the bottom, concrete being carried from about 1 foot beneath the sewer to an average of 4 feet above it. " The cost of the trench, including coffer-dam, sheeting left in place, and back-filling, was $44 per lineal foot." (Engineering News, vol. XXXI, page 121.) The material through which the trench was carried was sand and gravel. The work was done by day labor.

If the trench is in rock or a tight coffer-dam cannot be made except at great expense it may be cheaper and better to resort to divers. When not in rock, however, the excavation of the trench should be done by a dredge or similar appliance if possible, as divers' labor is very expensive.

The end of an outlet which discharges at some distance from the shore of a stream or other body of water should be so located and designed that currents, tides, or storms cannot wash it full of sand or mud, that it cannot settle down into the bottom, that it cannot be undermined by tides or currents, and that the sewage discharged will not settle in front of it and block the outlet. This may be accomplished by laying the sewer in a trench as described above, and at the very end placing a right-angled bend pointing upward and extending 1 to 3 feet above the bottom, this upright pipe being surrounded with a cone-shaped mass of concrete. Or the end of the sewer may be continued straight, but raised gradually until the outlet is 2 or 3 feet above the bottom, it being supported between two rows of piles back to where it has 2 or 3 feet of covering.

It is not so necessary that an outlet pipe be straight in line and grade provided the grade continually falls at a sufficient rate. The use of flexible-jointed iron pipe, such as is frequently employed for water-pipes at river-crossings, may often be used for sewer-outlets. They should be properly protected by concrete, riprap covering, or piling. For furnishing and laying 2200 feet of 24-inch iron pipe with Ward flexible joints in a bottom consisting of sand, gravel, loose and solid rock, in a trench having an average depth of 4 feet, the depth of water at high tide being $11\frac{1}{2}$ to 30 feet, from $16.85 per foot to almost double this amount was bid in 1898. In ordinary river-work the cost should be much less.

Whenever subaqueous work of any considerable extent is being done it will be well to have a diver's outfit on hand, as its immediate use may sometimes effect a saving of the work from serious damage.

ART. 80. CROSSING RAILROADS AND CANALS.

Railroads should be crossed with particular care, both that no accident may occur to either the workmen or to passing trains, and because of the difficulty of afterward repairing breaks or defects at such points. This applies also to sewers constructed in or close to the foot of railroad embankments.

It is not advisable to tunnel under a railroad unless the sewer runs quite deep and the material is stable. A settling of the ground above the tunnel might prove disastrous to trains, and this settling is extremely probable, owing to the jarring of passing trains. If there is a culvert under the road through which the sewer can be passed it will often be well to take advantage of this, if only a slight detour be necessary in order to do so.

If the sewer is to pass under the railroad in open cut each rail should be first supported by bridge timbers, beams, or iron rails placed under the ties lengthwise of the track and extending 10 to 20 feet beyond each side of the proposed trench. For a trench 4 to 6 feet wide a 12 × 12 bridge timber 25 or 30 feet long may be used. A heavy steel rail may be used under the same conditions, but is not generally so stiff. Each beam or rail is placed in a trench dug under the ends of the railroad-ties, just sufficiently deep and wide to enable it to be placed under the track-rail. Hardwood plank are then driven between this and the ground and wedges driven between each tie and the beam. The trench is then excavated, horizontal sheathing being used. The earth excavated cannot be thrown upon the surface unless the track is temporarily out of use. It may be handled by a cable-way excavator which swings sufficiently high to clear all trains. The buckets for this it will be well to have large— those holding a cubic yard will do—that the number of trips may be lessened. The back-filling can be returned in the same way and should be most thoroughly tamped.

Another method of handling the earth, particularly applicable where there are but two or three tracks, is to throw the excavated material beyond the outside track by one or two handlings, a space for this having been left clear of earth by previous management. If the trench is shallow and as short a length as practicable opened at a time it may even be possible to throw the excavated earth directly onto the completed sewer, but if this is done only a very few men can be worked at this point.

After the completion of the work with thoroughly tamped back-filling the trench should be wet down every two or three days for several weeks, the bridge timbers or rails being left under the ties meantime. Just before each wetting earth should be placed and tamped on the filled trench to 2 or 3 inches above the ties. When the trench shows no settlement after a wetting down the supporting timbers or rails may be removed.

For small sewers it will be well to use iron pipe with lead joints for railroad-crossings, and for large sewers the arch and side walls should be reinforced (see Plate VI, Fig. 8). In general it is better to place no manhole or other appurtenance between or within several feet of any tracks.

A trench in or near a railroad embankment is subject to the jarring of the trains and needs to be carefully sheathed. This is sometimes difficult if the trench be wholly or partly upon the slope of the embankment, since there is nothing opposite the upper ranger on the up-hill side against which to brace it. It will not usually be practicable to place a sloping brace from this to a lower ranger on the opposite side. A better plan would be to brace the sheathing against posts driven at intervals a little distance from the lower side of the trench and throw all the excavated material against this side.

The sheathing on the lower side at least should be left in and protruding a short distance above the ground after the work is completed, to prevent the back-filling, which should all be thoroughly hand-rammed, from being washed down the bank by rain.

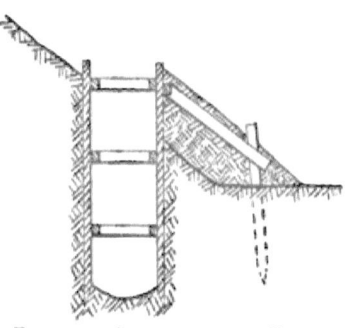

FIG. 32.—SHEATHING ON STEEP SLOPES.

It is not impossible to construct a sewer under a canal, raceway, or other body of water retained by embankments without drawing off the water or interfering with its service, but it is much easier and safer to do this work when, if ever, the water is out. The construction of a system can generally be so managed that all canal-crossings may be made in winter while the water is out, even if no other part of the system is constructed at that time. A raceway can in many cases be carried over the excavation temporarily by a flume extending for some distance in each direction from it. Care must be taken to prevent the water following the outside of this, for which purpose close sheet-piling may be used to advantage, being driven across the raceway at each end of the flume and making a tight joint with it.

A sewer under or near a canal should be of iron pipe, unless too large, when concrete may be used, made very strong and extra thick—say 1 part Portland cement, 2 parts sand, 3 parts broken stone, with a 50-per-cent increase in thickness over ordinary localities. If iron pipe be used cast-iron flanges made in halves should be bolted on the pipe at intervals, a thin lead strip being placed between the pipe and the flange casting to make a water-tight joint, or lead being calked into bells on the flange, as in the case of a sleeve-joint. Two or three of these flanges should be placed in each embankment and others 10 or 15 feet apart through the canal.

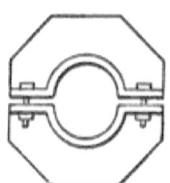

FIG. 33.—FLANGE FOR PIPE IN EMBANKMENT.

All space under, around, and above the pipe should be thoroughly filled with puddled clay, gravel, and sand carefully rammed. If clay cannot be had loam may be used, free from roots or "muck." A good proportion for these materials is 1 part of clay, 1½ parts of sand, and 4 parts of gravel, thoroughly mixed before placing in the trench. If the sewer is of concrete flanges of the same material may be built around the barrel at intervals; or the flanges may be of stone masonry, water-tight, with rough face. The flange, whether of iron, concrete, or stone, is better the rougher it is. It would be well to imbed rough stones in the entire outside of a concrete sewer under a canal to prevent the water following the surface and creating a leak.

If the earth over the sewer in the canal-bed is shallow or is not absolutely impervious there must be sufficient weight in or attached to the sewer to prevent it from floating if empty. A 24-inch iron-pipe crossing only two or three feet under a canal has been known to break in two at a joint and a part of it rise through the thin earth covering into the water above on account of the hydrostatic pressure brought to bear by seepage-water. It must be remembered that an empty iron pipe 36 inches diameter, for instance, to weigh as much as the displaced water must be 1⅜ inches thick. Consequently the heavier weights of iron pipe should be used, or else they should be weighted down with concrete, iron castings, or in some other way. It will usually be found cheaper to use the heavy pipe.

If it is necessary to pass a sewer under a body of water in tunnel this may require the use of compressed air, shields, etc., and should not be undertaken without the advice of an expert in such work.

PART III.

MAINTENANCE.

CHAPTER XIV.

HOUSE-CONNECTIONS AND -DRAINAGE.

ART. 81. NECESSITY FOR INTELLIGENT MAINTENANCE.

It is the too general rule that when a city has constructed a system of sewers it considers its duty done, and permits any kind of connection to be made with them, by anybody and in any way, and takes no more thought of its sewers until compelled to do so by some obnoxious conditions therein. This is all totally wrong, and even criminal. While it is not probable that any well-designed and constructed sewerage system will ever become "worse than no system at all" or an "elongated cesspool," it will not work at its best efficiency and free from objectionable conditions if unattended to, any more than would any mechanism.

Moreover, a considerable expense has been incurred to provide sanitary sewerage for the citizens, but if careless or penurious landlords or plumbers or ignorant householders are permitted to construct between the sewer and the house, or in the latter, cheap and unsanitary house-connections, -drains, and plumbing fixtures the health of the citizens is endangered and complete return for the outlay for sewers is not received. No dread of paternalism should interfere with the proper performance by the city of its manifest duty to require that all "sanitary" piping and fixtures throughout the city *are* sanitary, and the sewers should be in the charge of an experienced officer who is held responsible for their cleanliness and efficiency.

The first necessity for this oversight will come with the connection of the dwellings to the sewers.

ART. 82. REQUIREMENTS OF SANITARY HOUSE-DRAINAGE.

No house-connections should be attached to a sewer except in the presence and under the direction of a city inspector and by a party who is under bond to follow the city's regulations for such work.

No house should be allowed to connect with the sewer until its construction is entirely completed, including plastering and sanitary fixtures, owing to the danger that mortar and rubbish may otherwise be admitted to the sewer.

No connection should be made with a sewer except at a branch provided for that purpose. If there should be no branch within a short distance one may be inserted in a brick sewer by cutting through its wall and building a slant firmly in place or, in a pipe sewer, by removing a pipe and inserting a branch pipe in its place. If 3-foot lengths of pipe were laid in the sewer a few 3-foot lengths of branch pipes may be kept on hand for this purpose. (Branch pipes are generally used in 2-foot lengths.) To remove a pipe from a sewer it may be broken to pieces with a hammer, care being taken not to crack the adjacent pipe. Then, with a cold-chisel used with some care, the upper half of the bell facing this opening is broken away and likewise the upper half of the bell of the branch pipe to be inserted. This is then dropped into place with the branch on the wrong side and revolved, thus bringing to the top of the sewer that part of both pipes where the bell is wanting. The joint is then made, Portland cement being substituted for the missing portions of the bells.

In breaking the cap or plug out of a sealed branch care must be taken not to break any part of the pipe. If broken the pipe should be replaced by a new one, as above. If the branch is cracked it may be left in, but should be surrounded with rich cement concrete well compacted.

It is absolutely not permissible to cut a hole into a pipe sewer and insert the house-connection therein, as it is almost impossible to obtain a junction which will not leak or to prevent the connection-pipe from protruding into the sewer.

The house-connection should never be larger than the branch which it enters, but should preferably be smaller. A 4-inch pipe is large enough for any residence or small hotel or, in general, for 90 per cent of all the buildings in most cities. On a grade of 1 : 40 it should carry the simultaneous discharge of ten or more water-closet flushes, or that of two large bath-tubs when emptying themselves in two minutes. This connection may be of vitrified clay pipe from the sewer to a point 5 or 6 feet outside of the cellar wall. It should be laid to as perfect line and grade as was the sewer itself, the fall of 1 : 40 being the minimum allowed under any but exceptional circumstances. If a uniform grade from the sewer to inside the cellar is not obtainable or desirable, or if this distance be more than 100 feet, it is advisable to place an inspection-hole at the fence-line or at some other convenient point (see Plate XI, Fig. 10), the grade and line being straight each way from this to both sewer and house. If the pipe branches before reaching the house an inspection-hole should be placed at the junction. The joints of the house-connection should be of cement, and it should be of equally as good material as, and laid in every way according to the methods used for, the sewer. In made ground or quicksand, or where trees are near the pipe, or the latter passes near a well or cistern, the connection should be of iron water- or gas-pipe (not "plumbers' pipe") with lead joints.

From a point 5 or 6 feet outside the building into and through this the main pipe should be of iron, and should extend vertically to and through the roof, its upper end, down to a few feet below the roof, being preferably enlarged somewhat. The top should be at a distance from any chimney and above any garret or other windows, and should not be furnished with a cowl, quarter- or half-bend, or any other device. All fixtures in the house discharge into this pipe, the intersections being by means of Y's and not T's.

So far all authorities agree. But the general arrangement of traps and ventilation-pipes is a point upon which many of them differ. The principal point of difference is as to whether the pipe should be furnished with a trap between the sewer and the vertical "soil-pipe." Most agree that a trap should be placed just below each fixture, although a few would dispense with this and rely upon one main trap only. (See "The Single Trap System of House Drainage," Transactions Am. Soc. Civil Engineers, vol. XXV, page 394.) Some trap is desirable, if for no other reason than to prevent long sticks, bones, knives and forks, and other large articles from being carried to the sewer. Most sanitarians would ventilate each trap by connecting the end furthest from the inlet with a main vent-pipe leading through the roof.

The object of the main trap, which is generally placed just inside the cellar wall, is to exclude the air of the sewer from the building. As has been previously stated, however, the house-connection and soil pipes are in most cases much more foul than the sewer, and the danger lies in them rather than in the sewer. The vertical soil-pipe, on account of the spraying action of falling water, becomes fouler and is more difficult to clean than almost any other part of a sewerage system. Hence the author can see little if any advantage in the presence of the main trap; none, certainly, if this be not vented on both ends to prevent its seal being forced by a compression of the sewer-air due to a sudden discharge into the sewer from a near-by connection or some other cause, and to prevent a forcing of the traps throughout the house by the compression of air in the soil-pipe caused by a considerable flush of water from a fixture on the upper floors; also to admit fresh air to the house-piping. Many excellent authorities, however, advise the use of the main trap.

The object of continuing the soil-pipe through the roof is to allow the foul air from below to pass upward through it, and there seems to be little objection to permitting the purer air from the sewer to occasionally take the same course, and even perhaps some advantage from its diluting effect.

Whichever the plan adopted, if the workmanship and material are of the best there is probably little danger to be feared. If vent-pipes are used on a main trap these should terminate at a distance of at least 10 feet from any window or door, and in such a manner that they cannot be sealed by dirt, snow, ice, or frost collecting around the upper ends from the damp sewer-air.

"Every trap and dead-space in water-closets must be separately vented at the top of the outer bend, the branch vents connecting with the main vent," which should be carried from the lowest trap up through the roof. This prevents siphoning of the traps by water plunging down the soil-pipe from a higher closet or tub, and offers escape for any foul air forming in any of the soil-pipes.

Authorities agree that water-closets must not be connected directly with the water-supply pipes, but should be flushed through an intermediate tank or tanks or similar appliance; that roof-water leaders should never be connected directly with the house-connection pipe without an intermediate trap; and that all pipes and sanitary fixtures of whatever kind should be everywhere accessible for examination, and should not be walled or even boxed in. The water-closets should never be placed in rooms not receiving light and air directly from outdoors, or at the very least from a large air-shaft, through a window having at least 4 or 5 square feet area.

All piping in and near the house should be of iron. Wrought iron with screw-joints is preferable to cast iron. The use of lead pipe is not advisable, except in short exposed lengths, as it may be punctured by nails or gnawed by rats and mice, and is apt to sag into unnecessary running traps. Where the pipe passes through the cellar wall this should be arched, leaving a space of two or three inches around the pipe to prevent the breaking of the pipe by a settlement of either it or the wall.

The house-connection should be suspended upon the side walls after entering the cellar, and should never be placed under the cellar floor, unless, when this is unavoidable, it be placed in a shallow trench having brick or stone walls and with a removable top forming part of the cellar floor. The soil-pipe should have only easy curves and Y's, no angles or T's anywhere.

Water-closet tanks should discharge not less than 5 gallons with each flush, the pipes leading from these to the closets being not less than $1\frac{1}{4}$ or $1\frac{1}{2}$ inches. No overflow from any cistern, tank, or refrigerator should discharge into any soil- or waste-pipe, but into a trapped sink or bowl connected therewith, the end of the discharge-pipe being at least 3 inches above the water in said sink or bowl. There should be no wooden wash-tubs or sinks. Grease-traps, if used, should be cleaned out once a week. No "bell trap" or removable strainer should be placed in sinks or tubs. All iron pipes used as drains or soil-pipes should be coated inside and out with coal-tar varnish, or asphalt, or better still with enamel. A hand- and inspection-hole should be placed in the house-connection just inside the cellar wall, and outside the main trap if one be used.

Every system of house-drains and soil-pipes should be tested by water-pressure to at least 10 pounds before being accepted or used. (See also " House Drainage and Sanitary Plumbing," and "Sanitary Engineering," by Wm. P. Gerhard.)

To insure that the above requirements are met by every system of house-drainage—and this *should* be insured—regulations embodying them, and such others as are thought desirable, should be drawn up, and an inspector or inspectors appointed to examine and approve of all plans for house-drainage and to see that these are faithfully carried out.

CHAPTER XV.

SEWER MAINTENANCE.

ART. 83. REQUIREMENTS OF PROPER MAINTENANCE.

THE requirements for keeping a sewerage system in good running order can be concisely stated as—preventing and removing deposits, and maintaining ample and safe ventilation.

As previously stated, the main dependence for preventing deposits is flushing. If a deposit remains for any time it is apt to continually increase and become more difficult of removal, and deposits should therefore be removed as soon as possible after forming. This the automatic flush-tank is supposed to do for 800 to 1000 feet below it, but any forming below this limit will probably need to be removed by hand-flushing from a manhole or by the use of special appliances. If deposits continually form in any one place and are not apparently occasioned by articles which should not be introduced into the sewer it may be advisable to place a flush-tank at the head of where such deposits form, at one side of the sewer, but connected with it at a manhole or by a Y branch. If obstructions are frequently formed at any one place by the introduction of improper matters, such as ashes, bones, etc., the source of these should be ascertained and the parties responsible therefor punished.

It should not be taken for granted that a sewer is working properly, but the system should be inspected once a week or at least once a fortnight. This may require merely a look into each flush-tank to see that it works properly, into each inlet or catch-basin to see that it is clean and the grating unobstructed, and into each manhole (the dirt-pan being at the same time removed and emptied) to see that the sewage is flowing with sufficient velocity and is apparently not dammed back by any deposit below. But during the first few months of his service the inspector should enter each manhole and look through the sewer at each inspection until he becomes familiar with its condition of depth and velocity of flow when in good order. If there are any considerable odors observed about any appurtenance the cause should be discovered and removed. This will usually be a large deposit or imperfect ventilation, except in the case of catch-basins, where it probably means improper or infrequent cleaning.

The catch-basins should be cleaned after every rainfall. There is danger of putrefaction and objectionable odor from these if this is not done within two or three days after each rain, but this is almost impracticable in large cities, where there are one or two on every corner, without the use of an enormous number of men and carts, since each cart with three men will clean but five to ten catch-basins a day. As an example of what is usually done in this line, a large city in New England, which is considered to have an excellent Department of Public Works, during the entire year of 1890 cleaned its 1100 catch-basins an average of 1.84 times each. It seems almost impossible that these catch-basins could hold the heavier matter washed from the streets during six or seven months (or if so the small amount contributed by each storm would have done little harm in the sewer), and the inference is that a large part of this was not held, but was washed into the sewer; also that the catch-basins were in an unsanitary condition a large part of the time. When so treated they might better be replaced with plain inlets.

A record should be kept of all sewer-inspections, each line of sewer and each appurtenance having a record of its own showing when it was inspected, its condition, when cleaned, what repairs were made to it, with their nature and cost; of the frequency of flushing or of the discharge of each automatic flush-tank; of the location and date of making each house-connection, with all details as to route, size, and grade of connection-pipe, cost, by whom ordered, by whom put in (if by private contractor).

The house-drainage is usually supposed to be, but seldom is, looked after by the owner. It is exceedingly desirable to have a sanitary inspection made of every house by a city inspector at intervals of not more than 12 months; but such a plan would hardly be favored by most American communities, but would be looked upon as an impertinence. It is the city's duty, however, to insist upon all owners and tenants observing the sanitary regulations as to construction and use of house-drainage systems.

Extensions of the system should of course be made with as much care as were the original sewers, and no alterations made in the original plans without a careful consideration of their effect upon the system as a whole.

ART. 84. FLUSHING.

When automatic flush-tanks are used they should be inspected at intervals to insure their regular discharging. The most common failing with siphon-tanks is the trickling over of the water into the sewer as fast as it enters the tank after it has once reached the level of the top of the bend. Under this condition the siphon will never flush. This trickling may be due to faulty designing, but is usually caused by a leaking joint or blow-hole in the iron siphon at some point, which must be corrected. The frequency of discharge is regulated by the cock admitting the water. This can be adjusted only by actual trial with each tank. It is a good plan to have one or more registering reservoir-gauges for use in the flush-tanks which will indicate the times of discharge. A simple one, but sufficient for this purpose, can be made with a clock-works actuating a cylinder on which the height of water is constantly registered by a pen whose motion is caused by the rise and fall of a float, the pen and a rod from the float being attached to opposite ends of a lever with unequal arms, so that the path of the pen is but 4 or 5 inches long. Such an apparatus left for a day or two in a flush-tank will serve in place of frequent visits to it, and can be moved from one to another as each is adjusted to the desired frequency of discharge. The waste of water caused by flushing oftener than once in eighteen to twenty-four hours is not justified by any proportionate advantages.

Reference has already been made (Art. 47) to flushing directly from 2- or 4-inch branches led from the water-main into the flush-tank. In using these the valve is ordinarily opened to its full extent, or so much as is necessary to maintain the height of water in the flush-tank as great as is safe for the tank or sewer. It may be left open until such time as the water flowing through the manholes below is perfectly clean. It will be necessary to use the most solid construction in the flush-tank to resist the considerable force with which the water leaves the water-pipe.

Instead of connecting the flush-tank with the water-main by a large pipe a small one is sometimes used, and the tank filled from this after closing the sewer end, which is then opened and the contained water allowed to flush the sewer. This method takes much longer than the previous one and is consequently more expensive. In some cases the flush-tank is filled by hose from the nearest fire-hydrant.

In some cities the water is conveyed to the flush-tanks in

carts, and either the tanks filled from these and discharged by hand as above, or from the bottom of the cart a large pipe or canvas hose is lowered into the flush-tank and connected with the end of the sewer, into which the water is discharged under a head equal to the elevation of the cart above the sewer. In New Haven, Conn., such a cart is used holding 700 gallons, in connection with which an ovoid ball is passed down the sewer to assist in the cleansing, its distance from the flush-tank being regulated by an attached cord which passes up through the sewer and flushing-pipe to the surface. These carts are ordinarily used at manholes along the line of the sewer rather than at flush-tanks proper.

Flushing, as has been stated, is seldom effective for more than 800 to 1000 feet below the point of entrance of the flushing-water. Hence, when automatic tanks are not used at the head of every section of such length which requires flushing, this is performed at manholes wherever necessary. For this purpose outside water may be introduced by carts, as just described; or all the openings in a manhole may be stopped and the manhole filled by hose, when the plug to the down-stream opening is removed and the sewer below flushed; or only this opening is closed, and the sewage is permitted to back up in the sewer above, when the plug is removed and the sewage performs the flushing. The last method is not particularly satisfactory with pipe sewers in most instances, since the head obtainable is usually very small and the velocity of flush consequently the same, and if the house-connection pipes are on a flat grade the sewage may back up these to an undesirable height. Deposits also may form while the sewage is accumulating, which will not be removed by the flush if near the upper end of the dammed sewage, and the time required for a sufficient volume of sewage to collect will often be considerable and increases directly as the necessity for frequent flushing in each case.

The plugs used for stopping pipe and small brick sewers may have any of a variety of forms. One design is a simple conical cork-shaped piece of wood with heavy rubber so fastened around it as to come between it and the inside of the sewer when the plug is pushed into place and make a water-tight joint. Another consists of a solid centre of plank, around the edge of which is placed a pneumatic tube similar to a bicycle-tire, which is inserted just inside the sewer and the tire inflated by a bicycle-pump. These have ropes attached by which to draw them out of the sewer when the manhole or flush-tank is full, the air being first released from the tube of the one last described.

Another plan, that of bracing a loose frame or hinged gate against the end of the sewer in a manhole, is hardly applicable to properly constructed systems, where the manhole-channel and sewer are continuous, but may be used in a flush-tank designed for the purpose. The cover, whether loose or hinged, may be held in place by a brace hinged at the middle and extending from the cover across the flush-tank to the opposite wall. A rope is attached to the hinge of the brace and by pulling this when the tank is full the brace folds up and releases the cover.

In large sewers it is generally impracticable and unnecessary to dam back the sewage higher than, or even as high as, the crown of the sewer, and a dam one half or two thirds the height of the sewer is sufficient. This may be made similar to those already described, but not filling the entire bore of the sewer. Or a "pocket dam" may be used. This consists of a bag of tarred canvas having rings around its mouth and a rope passing through these long enough to reach from the sewer to the surface. Another rope is fastened to the bottom of the bag. This bag is filled with water and placed in the sewer-invert, being held upright by the rope through the rings, and serves as a dam to the sewage. When this has raised sufficiently this rope is released, the bag collapses and is removed by the rope attached to its bottom.

In very large sewers flushing, if practised at all, must generally be done with sewage, on account of the enormous quantity of water required for this purpose. But this practice is not recommended where sufficient water can be obtained. In the case of storm or combined sewers advantage should be taken of light rains by damming up the run-off from them in the sewers and flushing with this comparatively clean water. Heavy storms of course need no assistance in their flushing effect.

To ascertain the height to which water in a large sewer has risen in flushing (or at any other time, as during storms) an ingenious method, employed at Omaha, Neb., is to drive into the wall, 2 inches apart vertically, small iron rods with the ends turned up, on each of which rests a cork with a hole in its bottom, which can be readily floated off when reached by the water. Upright whitewashed sticks placed in the vertical diameter of the sewer have been used for the same purpose, but not with perfect success.

Of the above methods of flushing Andrew Rosewater considers the automatic flush-tank the least expensive, the use of 4-inch water-pipes with hand-valves next, then the use of hose from the hydrants, and the water-cart method the most expensive. Cleaning sewers in New Haven by the water-cart above described cost $3 to $4 per mile cleaned. One argument in favor of hand-flushing is that it renders more probable frequent inspection of the system, which will be made at the time of flushing; but on the other hand pressure of other duties or carelessness may cause longer intervals between flushings than is desirable. As a general rule automatic tanks should be used on pipe sewers where there is not retained by the city a constant force of laborers for maintenance of sewers and streets and similar purposes. In the case of large brick sewers it is probably best to resort to one of the methods of hand-flushing. For pipe-sewer dead-ends in cities with a maintenance force automatic appliances are desirable, but are in many instances not used. When any flushing is done elsewhere than at dead-ends hand-flushing is generally resorted to.

Art. 85. Cleaning.

The purpose of flushing is to prevent deposits, or rather to prevent the accumulation and solidifying of deposits. But from the insufficiency or infrequency of flushing this object is sometimes not attained; or obstinate obstructions may be formed by sticks, stones, or other matter which flushing is not expected to remove, and these must be removed by hand or some other method. Catch-basins must be cleaned by hand, and this should be done frequently. The manhole dirt-buckets, also, should be cleaned at intervals. These last are merely removed from the manholes and dumped into a cart or wheelbarrow.

The catch-basins are generally cleaned by ordinary shovels, the dirt being taken to the surface by a bucket and emptied into a cart. Two men and a cart and horse suffice for this work. In some cities, and especially when the catch-basins are small, the dirt is removed with long- and heavy-handled hoes, the blade of the hoe being at right angles to the handle and about 8 by 10 inches in size. These are used from the surface through the manhole-opening or that left by removing the grating. Catch-basin walls should be thoroughly cleaned with a hose and broom and washed with a solution of chloride of lime or some deodorizer, but this is seldom done. The cost of cleaning a catch-basin will vary probably from 50 cents to $2 each, depending upon their size, the frequency of cleaning, and other special circumstances or conditions; $1.40

seems to be about the average for large cities. Catch-basins at the ends of siphons are difficult to clean, being in most cases at the bottom of a shaft containing many feet of water. Long-handled hoes may be used, or the siphon may be closed and emptied of sewage to permit reaching the catch-basin. An apparatus acting on the principle of the steam-siphon or sand-pump is used with success in the Waltham, Mass., siphon, emptying the catch-basin or sump without the siphon being emptied. The pipe B, Fig. 34, is lowered into the sump and the nozzle is attached to a hose from a hydrant. When the water is turned on the sand and other solid material, mixed with sewage, is sucked up through B and discharged through A into the sewer, from which it is prevented from returning by a temporary dam in the end of the sewer.

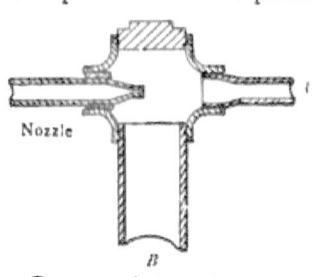

FIG. 34.—APPLIANCE FOR CLEANING SIPHON-SUMP.

Small sewers are cleaned by flushing when this is possible, but in many cases other means must be resorted to. The use of "pills" is convenient where there are no stones, sticks, or other hard materials in the sewer. These are round balls, usually of wood, which are floated through the sewer either in the sewage or, if there is not enough of this, by flushing water. A set of these 2, 3, 4, 5, 7, 9, etc., inches in diameter should be kept on hand. When a sewer is to be cleaned the smallest pill is floated through from one manhole to the next, where it is caught by an assistant; the others are then sent through in the order of their sizes until all have passed through up to the size one inch smaller than the sewer. When any ball reaches a point where the opening is contracted by sediment to less than its diameter the ball, which has floated and rolled along the top of the sewer, dams up the water until it has sufficient head to force its way under the ball and scour out the sediment. The ball rolls slowly ahead, the current washing away the sediment for an inch or two under it. If there is a lamp-hole on the line the ball may bob up into it, and a man should be stationed there with a pole to push the ball down and into the sewer below the lamp-hole. If a stone or stick is among the deposit the ball may be stopped by it, in which case both stone and ball must be removed by another method. The pill cannot be used when the sewer is stopped entirely so that there is no flow through it. No cord should be fastened to any of these round balls, as it is liable to be rolled about them and wedge them in the sewer, catch in obstructions, and generally give trouble. Ovoid balls, however, are sometimes used with cords attached. These do not roll along the top of the sewer, and may need to be weighted to prevent the friction between them and the sewer top interfering with their motion ahead.

In place of the pill, particularly in sewers larger than 12 or 15 inches, a small carriage is sometimes used which travels on wheels through the sewer, its front being of such a shape as to almost fill its bore except for an inch or two at the bottom. Where the sewer is not more than 3 or 4 feet in diameter the carriage is usually provided with other wheels on top, which are pressed against the sewer-arch by springs. This contrivance is hauled through the sewer by a rope, which has first been introduced into it by floating through the sewer, a piece of wood or cork carrying a cord to the end of which the rope is attached. Another rope is fastened to the rear of the carriage to haul it back if it strikes an immovable obstruction. This is a modification, and on a small scale, of the method employed for cleaning the Paris sewers, where a plank form, similar in shape to and but little smaller than the sewer-invert, is carried by a boat or wagon and lowered into the sewer as far as necessary to cause a scouring of the deposit. The boat or car is carried forward by the water backed up behind the scouring-form, which is raised or lowered to the proper position by a workman riding in the conveyance.

These methods all depend upon the scouring action of the water and presuppose a passage through the sewer. Other contrivances for cleaning a small sewer under such circum-

FIG. 35.—DISK FOR CLEANING SEWERS.

stances are based upon the use of main strength to haul the material out. Probably the simplest is in the shape of a heavy plank disk to which a rope is attached by three short light chains fastened to as many bolts through the disk. One of these chains is attached at each side and one at the bottom of the disk, and their relative lengths are so arranged that when all are taut the top of the disk will incline a little away from the rope. Upon the other side of the disk, at its top, is fastened another rope. By the latter it is pulled a short distance into the sewer, lying flat; the other rope is then pulled, when the disk rises into an upright position and scrapes along the deposit in front of it. It is well not to draw this too far into the sewer at once, but to clean only a few feet at each trip. The dirt can be scraped to a manhole and there removed by buckets. It is awkward pulling in a manhole bottom, and it is well to arrange a pulley in a frame, around which the rope passes, as also around another pulley at the top to permit of a horizontal pull. The lower frame may consist of two 4×6 or 4×8 timbers fastened to each other parallel and a short distance apart, between which the pulley turns in journals fastened to their under sides, these timbers being braced against the inside arch of the sewer and the pulley being in the centre of the manhole (see Fig. 36). This method can be used where the material is too heavy to be scoured out by pills or similar contrivances, and also as a substitute for these.

In some cases the sewer will be found entirely stopped, so that no cord can be got through it, and an opening must be forced through. A rod of some kind is used for this purpose. Since none longer than 5 feet can be got into the sewer through the manhole (unless it be too flexible for efficient service) rods of this length made to joint together are gen-

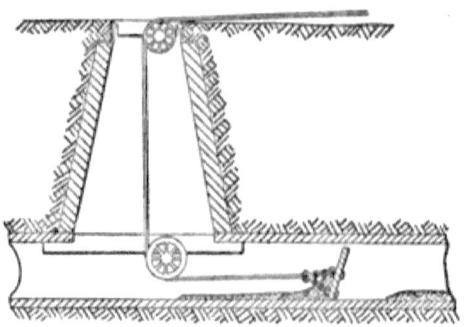

FIG. 36.—METHOD OF USING CLEANING-DISK.

erally used. These are sometimes lengths of gas-pipe with screw-couplings, or in some cities 1- to 1½-inch maple rods with brass screw-caps fastened to their ends are used. These are forced through the obstruction by working them back and forth or even by driving with a hammer. When an opening is once made it is well to leave the rod in it and work it a little back and forth as the sewage flows through until the hole is too large to be in danger of immediately stopping again, when a pill or cord may be floated through and the cleaning completed by one of the above methods.

A small sewer or sub-drain may also be cleaned by the use of hose, as explained in Art. 78.

In some cases the obstruction may be so obstinate as to necessitate the digging up of the sewer. Before doing this its exact location should be ascertained by pushing a rod to it through the sewer and measuring its length, or by the use of mirrors, as previously described.

For cleaning house-connections, sub-drains, and other small pipe which cannot be readily reached the hose is excellent, sufficient water being turned through it to make it stiff enough to be pushed through the pipe; or rods may be used, as for the larger sewers. Instead of a rod the city of Waltham, Mass., has used for these cases a length of steam-hose filled with sand, a wooden plug being fastened in the end of it. This is flexible, but stiff enough for use in a pipe only 3 to 5 inches in diameter.

Even pipe sewers of 18 inches diameter and up can be entered for inspection and cleaning by hand. It is reported that in Waltham a Hungarian crawled through 850 feet of 15-inch pipe running $2\frac{1}{2}$ to 4 inches deep with sewage, there being in at least one place not over 9 inches of clear space above the deposits and sewage. The author has seen a contractor crawl through 200 feet of 18-inch sewer, and it is nothing unusual for a man to pass through almost any length of 24-inch pipe. A large stone or a stick wedged across the sewer can frequently be removed in this way and the necessity for digging up the pipe avoided.

If the sewer is found to be broken in any place there is generally but one thing to do, to dig down to and replace it. A sewer which is only cracked or is leaking badly has been repaired by inserting inside of it a line of screw-joint pipe as large as can be slipped into it, and sealing the space between the two at the ends with cement. The substitution of new pipe would probably be cheaper in most cases, however.

When small pipe is only coated or contains but little deposit it is sometimes cleaned by the use of a wire brush, just the size of the sewer, fixed upon the end of a rod similar to those already described.

The cleaning of sewers large enough to permit a man to

work in them needs no special discussion. If they are large enough the dirt may be carried to the manhole in a low car running on the sewer bottom. In smaller sewers it may be shovelled or hoed into a pile at each of two manholes from a point midway between them and removed in buckets.

An inverted siphon may be cleaned as an ordinary sewer, after the sewage flow has been diverted to the other siphon-pipe or dammed up and the sewage contained in it pumped out.

In 1891 the cleaning of 123 miles of sewers in St. Paul cost $6208, or about $50 per mile; labor 20 cents and team and driver 30 cents per hour, foreman $4.18 daily. The cost of removing stoppages from small sewers will probably average about $2 or $3 each. The annual cost per mile of keeping a system of pipe sewers clean probably varies between $10 and $75 in most cases; it should not exceed $10 to $25 for a well-designed and -constructed system containing 20 miles or more of sewers, with intelligent, economic maintenance, and during some years no expense for this purpose may be required.

INDEX.

	PAGE
Acceptance of a system, Requirements for	215
Adams Sewage-lift	152
Advertising contracts	239
Air, sewer, Character of	96, 99
Alignment, Giving trench	244
Alleys, Sewers in	118
Angles and bends in sewers	72, 119
Arches, Providing for thrust of	298, 308
Assessments, Conditions governing	233, 236
" , Methods of making	231
Atlantic City, Gaugings of sewage flow at	39
Back-filling, Cost of tamping	212
" " , Laborers used for	268
" " , Specifications for	212
Barrel, sewer, Shape of	81, 158
Beginning construction, Points for	241
Bends in sewer lines	72, 119
Benefits derived from sewerage	1, 2, 18, 139, 236
Berlier system of sewerage	7
Bids, Receiving and considering	239
Blacksmith, Necessity for, during construction	266
Borings, Test, along sewer lines	109
Boulders, Removing, from trench	274
Braces, Extensible	287
" , Measuring trench for	287
Branches, Laying sewer	209, 242, 251, 296, 341
Brick for sewers, Cost of	227
" " " , Specifications for	195
Brick sewers, Abrasion of	77
" " , Centre for	300
" " , Cost of construction	230
" " in quicksand	322
" " , Invert backing for	297, 304
" " , Method of building	299, 316
" " , Templet for	248, 298

	PAGE
Bridging trenches, Specifications for	201
Burlington, Vt., Gaugings of sewage flow at	40
C, Effect of variations in R upon	62, 71
", Meaning of	61
", value of, Formulas for the	62
Canals, Crossing under	337
Capacity, Designing for future	113
Catch-basins, Cleaning	147, 347, 354
" ", Construction of	181
" ", Where to use	146, 347
Caving of banks, Avoiding	273, 279
Cement, Cost of	231
" pipe joints, Cost of	228
" sewer-pipe, Use of	158, 170
" " ", Specifications for	194
" , Specifications for	197
Centre for brick sewers	302
Cesspools	3
Chemical precipitation	26
Chézy formula	61
Circular sewers, Conditions favoring use of	81
Cleaning large sewers	356, 360
" sewers, Cost of	360
" small sewers	264, 326, 355
" up streets, Specifications for	215
Coffer-dams, Construction of	332
Combined system defined	9
" " vs. separate	10
Concrete sewers	163
" ", Method of constructing	304
" ", Specifications for	203
Consumption of water, Daily and hourly variation in	33
" " ", Estimating future	33
" " " per capita, Table of	32
Contract, Form of	223
Contracting work, Advantages and disadvantages of	237
Contractor, Duties of	220
Contracts, Advertising	239
" , Awarding	240
Cost, Relative, of sewers of different capacities	57
" . See material or work in question.	
Cremation of sewage	28
Cross-section of sewer, Effect of shape of, upon velocity	69, 81
Curves, Loss of head in	72
Cutting sewer-pipe	297
Dead-ends in sewers	119
Defective sewers, Contractor's responsibility for	215, 217

	PAGE
Delays of construction, Provision in contract for	217
Depth of sewage, Minimum permissible	77, 83
" " sewer desirable	132, 148
" " ", Relation of Q, S, and d to	131
Design, Data necessary for the	103, 106
" , Principles of sewerage	111, 136
Des Moines, Gauging of sewage flow at	41
Dilution, Amount of, necessary for discharge into streams	24, 114
" , Disposal by	20
" " , Conditions affecting	24, 28, 114
" in tidal waters, Requirements for	25, 108, 114, 334
" , Purification by	21
Discharge into rivers and tidal waters. See Dilution.	
" through circular sewers, Effect of depth upon	69
" " egg-shaped sewers, Effect of depth upon	70
" " sewers, Table of	64
" " " partly full, Calculating the	71
Disk for cleaning small sewers	357
Disposal, Aims of	16
" , Commercial aspect of	17
" defined	14
" , Laws affecting	12, 16, 19, 20, 24
" , Principles involved in	18
" , Treatises on	14
Drainage area, Ascertaining size of	103, 106
" " , Data necessary concerning	104
" districts	117
" of wet soils	139, 315, 317
Draining Trenches. See Water, Ground.	
Driving-cap for sheathing	285
Dry-sewage methods	2
Dwelling, Average number of persons in a	35
Earth-closet system	5
Earth, Sewage treatment by use of	27
Egg-shaped sewers, Proportional dimensions of	83
" " " , Advantages of, over circular	83
Engineer, Power of, in contract work	220
Engineer's duties before construction	241
" " during "	254
Estimate of cost, Data for making	226
Excavated materials, Classification of	198
" " , Placing of, on streets	270
Excavating deep trenches	271, 279
" machinery, Advantages of using	199, 249, 275, 290, 336
" " , Different kinds of	276
" " , Economy of	278
" trenches by hand	270

	PAGE
Excavating trenches, Cost of	229, 315
" " , Specifications for	198
Extra work, Specifications for	218
Factories, Amount of sewage from	35
Family, Average number of persons in a	35
Field-book, Form of notes in	250, 259
Filtration, Treatment by	27
Final estimate book, Method of keeping	257
" " , Definition of	256
" " , Preparing	260
" inspection	215, 260
Fire, Treatment by	28
Fish, Effect of sewage upon	21
Flat-bottom sewers	161
Floats, Use of	108
Flow in sewers, Theory of	60
Flushing, Appliances for	93, 179, 349
" by hand, Methods of	91, 350
" " " ", Relative cost of different	353
" " " vs. automatic flushing	353
" " roof-water	92
" , Efficiency of	90
" from streams and tide-waters	92
" " water-mains direct	95, 350
" , Intervals between	86
" , Necessity for	85, 97, 347
" , Proper methods of	88
" , Sea-water for	93
" , Separate sewers without	90
" water, Amount of, necessary	87
Flush-tanks, Amount of water from	37, 94
" " , Automatic apparatus for	93, 179, 349
" " , Construction of	178, 211, 308, 350
" " , Inspection of	349
" " , Locating	145, 347
" " , Method of building	308
" " , Specifications for	211
" " , Testing	261
Foremen, Number and character of	266
Foundations, Forms of	298, 309
" , Materials for	310
" , , Specifications for	202
" , Where needed	189
Gangs, labor, Size and number of	265, 267
Gorged sewers, Relieving	153, 185
Grade cord, Setting and using	245, 246
" rod, Form and use of	248

INDEX. 365

	PAGE
Grade stakes, Use of	245, 250
Grades of combined sewers	76
" " house-sewers	75, 134
" " sewers, Calculating	134
" " " , Desirable	134
" " " , Maximum	77
" " storm-sewers	76, 134

Ground-water. See Water, Ground.

Hose for cleaning small sewers	264, 326
House-connections, Capacity of four-inch	342
" " , Interference of storm-sewers with	137
" " , Junction of, with sewers	142
" " , Line and grade of	142
" " , Locating	242, 251
" " , Method of cleaning	359
" " , Necessity of careful construction of	100, 141, 340
" " , Regulations for	341, 346, 349
" " , Sewer air in	96
" " , Size of	79, 342
" " , Ventilating sewers through	100, 344
" " with deep sewers	188
" sewage, Amount of	31, 113, 120
" " , Gaugings of	37
Hydraulic radius, Definition of	61
" " , Formula for, in circular sewers	75
" " , Tables of	69
Ice, Danger in sewage-polluted	24
Imperfect work, Contractor to repair	217
Imperviousness of ground and run-off	51, 124
" " " , Determining	124
Incineration, Treatment by	28
Injuries, Responsibility of contractor for	217
Inlet connections	143, 181
" " , Ventilating sewers through	100
Inlets, Construction of	180, 308
" , Locating	112, 145
" , Specifications for	211
Inspecting sewers	260
Inspection-hole	342
Inspection of house-drainage	346, 349
" " sewers, Necessity for	252, 347
Inspector, Duties of	252, 255
Intersections of sewers	165, 215
Intercepting-sewers	116, 153
Interceptors	154
" , Leaping weir	183
" , Diverting	183

	PAGE
Invert backing	297, 304
" blocks	163
" , Definition of	245
" former for small sewers	263
Inverted siphons, Construction of	186
" " , Principles of design of	138, 185
" " , Where used	81, 132, 251, 329
Inverts for brick sewers	164
Iron castings, Specifications for	195
" , wrought, " "	196
Irrigation, Disposal by	27
Joint packing, Specifications for	197
Joints, Pipe sewer	118, 168, 204, 319, 329
" " " , Concrete around	295, 319
" , Ward flexible, Cost of laying	334
Kalamazoo, Gaugings of sewage flow at	40
Kuichling's laws of run-off	48
Kutter's formula	62
Laborers, Housing non-resident	269
Lamp-holes, Construction of	178
" " , Where used	145, 251
Laying sewer-pipe, Cost of	229
" " " , Specifications for	206
Leaks in sewers, Stopping	206, 261, 319
Levelling necessary for designing	105, 107, 242
Liernur system of sewerage	7
Lifting sewage, Methods and apparatus for	149, 150
" " , When necessary	148
" stations, Location of	152
" " , Number of, desirable	150
Lines, Locating sewer	117, 244
Manhole bottoms	176, 306
" buckets	178
" steps	172, 307
" tops	177, 249, 307
" walls	177, 306
Manholes, Building, in quicksand	327
" , Cost of	230
" , Crossing	174
" , Dimensions of	172
" , Drop	174
" , Location of	144, 171, 336
" , Materials and shapes of	306
" , Method of building	306
" on large sewers	176
" , Purposes of	100, 144
" , Shallow	172, 308

INDEX.

	PAGE
Manholes, Specifications for	210
" , Sub-drain	174, 327
Map required for designing	103, 107
Masonry, brick, Specifications for	205
" sewers, Materials and shapes of	297
" , stone block, Specifications for	206
" stone, Specifications for	204
" work in winter	296, 308
Masons, Number of, required	267
Materials of sewer construction. (See also material or appurtenances in question.)	157, 161
Maul for driving sheathing	285
Measurement of work	220, 256
Memphis, Gaugings of sewage flow at	40
Mirrors for inspecting sewers	262
Monthly estimates, Preparing	258
Mortar, Method of making	204, 300
" and brick, Handling	300
n in Kutter's formula, Values for	63
Notes of the work	250, 255, 257, 258
"Nuisance" defined	19
Object of a sewerage system	30
Obstructions, Causes of, in sewers	85
" , Passing, by siphon	251
Office buildings, Amount of sewage from	35
Old sewers, Using, in new system	155
Outlet, Deer Island, Cost of	333
Outlets for sewerage systems	20, 105, 108, 133, 148, 153, 334
" , Construction of	333
Overflows	185
Oysters, Typhoid fever germs in	21
Packing, joint, Specifications for	197
Pail for earth-closet or pail system	5
Pail system	4, 27
Paving, restoring, Specifications for	214
Payments, Times of making	223, 269
Picks	273
Piles, Methods of driving	309
Pills, Use of, in cleaning sewers	264, 355
Pipe, broken, Replacing, in sewers	341, 359
" , cement, Specifications for	194
" , " vs. vitrified clay	170
" , drain, Specifications for	194
" , " , Cost of	228
" , heavy, Methods of laying	292
" , House-drain	345
" , iron, Cost of	228

	PAGE
Pipe layers, Number of, required	267
" , sewer, Cutting	297
" , " , Price of	227
" , " , Strength of	167
" , " , Thickness of	166
" sewers, Cleaning 264, 325, 355	
" " , Cost of laying	229
" " , Imperfections common in	263
" " in quicksand	322
" " " wet trenches 318, 325	
" " , Inspecting 254, 262	
" " , Laying, up or down hill 291, 317	
" " , Methods of laying 291, 319	
" " , Specifications for laying	206
" , two-foot vs. three-foot lengths	170
" , vitrified, Specifications for	192
Pipes and conduits, Interference with, by contractor	200
" , Water, gas, or drain, in the trench	274
Plans, sewerage, Data required for	103
Platforms in deep trenches	271
Plumbing, Regulations for sanitary	341
Pneumatic systems	7
Population, Distribution of, in families and dwellings	34
" , Districts based on density of 36, 116	
" , Estimating future increase in 35, 113	
" per acre, Rule for calculating	36
Private property, Sewers on	120
Privies	2
Profiles, Information to be shown on 131, 137	
" necessary	105
" , Preparing 107, 131	
Providence, Gaugings of sewage flow at	38
Pumping, hand, Methods of 311, 316	
" , steam, " "	312
" . See Lifting.	
Pumps, Hand, on construction work	311
" , Steam, on construction work	312
" , sewage, Capacity of, necessary	149
" , " , Kinds of, in use	151
Quicksand, Building brick sewers in 249, 322	
" , " manholes in	327
" , " pipe sewers in 249, 322	
" , Detecting presence of	110
" , Handling trenches in 268, 317, 320, 322	
" , Qualities of	320
" , Removing, from pipe sewers	325
" , Sheathing in 281, 283, 321	

INDEX. 369

	PAGE
R. See Hydraulic radius.	
Raceways, Method of crossing under	337
Railroad crossings, Construction at	335
" " , Specifications for	200
Railroads, Sewers near	336
Rainfall, Rates of, in various sections	45, 123
Ransome method of building sewers	305
Relief sewer, Purposes of	185
Removing a pipe from a sewer	341
Rivers, Methods of crossing	328
" , Discharging sewage into. See Dilution.	
Rock, Determining presence of	109
" excavation, Cost of	229
" " , Measuring	257
" " , Sewers in	189
" " , Specifications for	201
Rods for cleaning small sewers	358
Run-off at Nagpoor reservoir	51
" " Washington, D. C.	48
" , Comparison of formulas for	55
" conducted through gutters	59, 112
" , Diagrams for calculating	47
" , Factors of	44
" , Formulas for	52
" , " " , Discussion of	54
" , Gaugings of, at New Orleans	47
" , Kuichling's laws of	48
" , Method of calculating	124
" , Roe's table for	54
S, Definition of	61
St. Louis, Gaugings of sewage flow at	40
Sand, Cost of	231
" for mortar, Specifications for	196
Sanitary sewerage, Requirements of	1
Schenectady, Gaugings of sewage flow at	39
Sections of sewers	158
Separate system vs. combined	10
" " defined	9
Septic tank treatment	27
Sewage, Causes of danger from	3, 18, 23, 25, 96
" , Composition of	15, 31
" defined	15
" , Value of	17
" . See House- or Storm-sewage.	
Sewerage, Arguments in favor of	1, 2, 18, 139, 236
Sewer-pipe. See Pipe, Sewer.	
Shape of sewer section	81, 158

	PAGE
Sheathers, Number of, necessary	266, 268
Sheathing, Driving-cap and maul for	282, 285
", Horizontal	281, 285
", Materials and dimensions of	284, 286
", Removing	289
", Skeleton	280, 286
" trenches, Cost of	229
" ", Methods of	280, 287, 321
" " on steep slopes	337
" ", Specifications for	200
" ", When necessary	273, 279
", When to be left in	288
Shone Ejectors	7, 152
Shoring buildings	200, 290
Shovels	273
Sidewalks, Sewers under	118
Silt-basins, Use and construction of	182
Siphons, Inverted. See Inverted siphons.	
Sizes of house sewers, Method of determining	31, 75, 78, 120
" " " ", Minimum	79
", sewers, Calculating, from sewage volume	120, 134
" " storm sewers, Method of determining	44, 75, 122
" " " ", Minimum	80
Specifications, Classification of	190
", Definition of	190
", Requirements of	191
" . See the material or work in question.	
Staging in trenches, Construction of	271
Steep slopes, Sheathing trenches on	337
Stone, masonry, Specifications for	195
", paving, " "	195
Stoneware sewer-pipe. See Pipe, Vitrified.	
Stores, Amount of sewage from	35
Storm overflows	154
" sewage, Data for determining volume of	122
" sewerage, Extent of	112
" sewer, Determining size of	57, 66, 80, 134
" water, Amount of, to be provided for	56, 58
Storms, Damage done by	57
" , First, second, and third class	56
Street surfaces, Breaking, for trenches	270
" ", restoring, Specifications for	214
Sub-drain pipe, Specifications for	194
Sub-drains, Cleaning	325
" ", Construction of	187
" " for handling ground-water	312, 317, 318
" " in quicksand	324, 325

	PAGE
Sub-drains, Inspecting	264
" " , Specifications for	209
" " , Necessity for	139
" " , Outlets for	140
" " , Size of	141
Sub-invert spaces	189
Surveys necessary for designing	106
System of sewerage, Which, to adopt	12, 115, 149
Tamping trenches, Cost of	212
" " , Methods of	213, 295, 297
" " , Specifications for	213
Templet for sewers	298
" , Inspector's or skeleton	262
Tidal reservoirs	148
Timber, Specifications for	197
Time-keeper, Duties of	266
Toronto, Gaugings of sewage flow at	39
Traps, Location and use of	97, 182, 343
Treatment, Sewage, defined	14
" " , Difficulty of	15
" " , Method of, to be adopted	115
" " , Methods of	26
Trench machine. See Excavating machine.	
Trench, Storm- and house-sewer in the same	120
Trenches, Excavating	270
" , Giving line for	244, 270
Tunnelling trenches	272
Typhoid-fever germs in oysters	21
Velocity in sewers, Effect of depth upon	69
" " " , Formula for	61
" " " , Table of	64
" , Maximum, permissible	77
" , Minimum, permissible in storm and combined sewers	74, 76
" , " " " house-sewers	71, 75
" , Uniform, in a system	135
Ventilation of sewers, Methods recommended for	101, 344
" " " , Necessity for	95, 98
" " " , Various expedients for	98
Vitrified clay pipe. See Pipe, Vitrified.	
Walls, Thickness of sewer	159, 165
Water-carriage system	7
" closet tanks	344, 345
" closets, Location of	345
Water consumption and sewage flow	31
" " " , Estimating future	33, 114
" " " in various cities	32
" , ground, Amount of, leaking into sewers	37, 169

					PAGE
Water, ground, Detecting presence of					110
"	"	, Driven wells for lowering			318
"	"	, Handling		241, 310, 315,	318
"	"	, Sub-drains for handling			312
"	in trenches, Specifications governing				199
"	, Methods of constructing sewers under				329
"	, Tamping trenches with				214
Weston, W. Va., Gaugings of sewage flow at					40
Wet and quicksand trenches, Pipe joints in					319
"	"	"	"	, Size of gangs for	268
"	"	"	"	. See Water, Ground.	
Winter, Masonry work in					296, 308
"	, Pipe-laying in				308
Wooden-stave pipe					158
Working gangs, Size and number of					266

FOR LOW-LYING LAND, FLAT DISTRICTS, DEEP BASEMENTS

Adams' Automatic Sewage Lift.

(Without engines, pumps, or compressors.)

Raises *sewage* from low level *with sewage* from high level. In flat districts *with city water supply* at a cost—for producing the dynamic head in the required volume—of about 50 cents per 100,000 gallons of sewage raised.

Cost of sewerage systems reduced, river pollution avoided, cesspools avoided. Reference to users on application.

ADAMS' PATENT SEWAGE LIFT CO.,
623 Drexel Building, Philadelphia, Pa.
ALSO LONDON, GLASGOW, DUBLIN.

A. PRESCOTT FOLWELL, M. Am. Soc. C. E.,

EASTON, PA.,

CONSULTING ENGINEER FOR SEWERAGE, DRAINAGE, WATER SUPPLY, AND GENERAL MUNICIPAL WORK.

Designs furnished and Construction superintended. Reports on and Improvements and Extensions of Existing Systems.

THE MILLER AUTOMATIC SIPHON.

Pacific Flush Tank Co.,

84 La Salle St.,   Chicago, Ills.

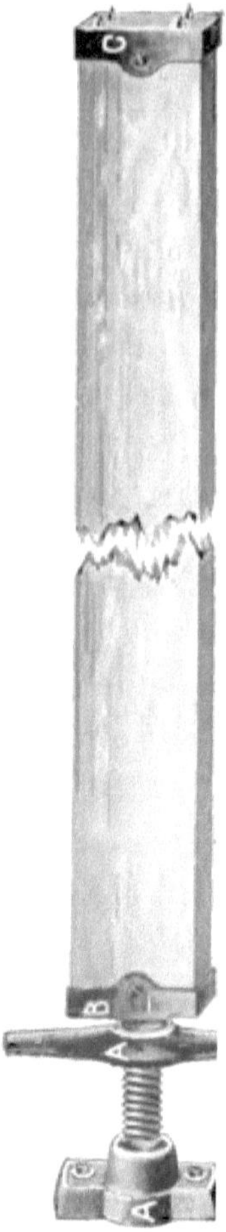

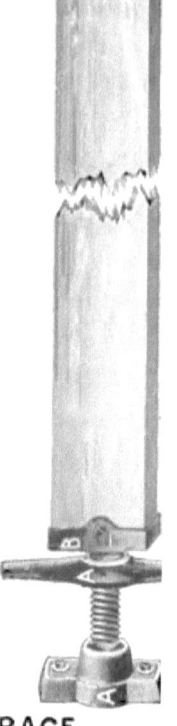

COMBINED SCREW AND TIMBER BRACE.
(PATENTED.)

Invaluable in wide and deep trenches where heavy bracing is required. Has given excellent results everywhere. Made with 1½" and 2" screws, and caps to suit all ordinary size timbers.

ALL-IRON TRENCH BRACE.

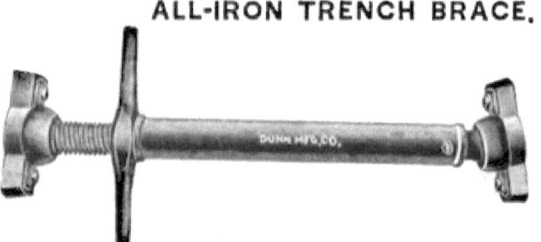

Suitable for all ordinary work. Strong and simple in construction. Made in all sizes. Endorsed by civil engineers.

Send for catalogue and price list.

DUNN MANUFACTURING CO., Pittsburgh, Pa., U. S. A.

ENGINEERING NEWS.

ESTABLISHED 1874.

Published Every Thursday.   Subscription, $5 a year.

"Engineering News" makes a special feature of important original articles relating to Municipal and Civil Engineering. It aims to record everything of importance in all branches of American engineering practice, together with such foreign matters as are most likely to interest engineers in this country.

The eight to twelve pages of Construction News items and Proposal Advertisements in "Engineering News" are read each week by prominent contractors and manufacturers of contractors' supplies in every part of North America, thus insuring an interest in work reported or advertised in this paper which often results in large savings on contracts.

We also publish books on engineering and allied subjects. We have specifications for viaducts and bridges of various kinds, steel roofs, buildings, structural steel, grading and masonry, cross-ties, track-laying, dams and reservoirs, wiring for electric light in buildings, etc. There are some thirty different specifications, the price being from 5 to 40 cents each. The following books are of special interest to municipal engineers.

	PRICE.
COST OF LAYING WATER PIPE. By C. D. Barstow, C.E. Pamphlet, 5 x 7 inches, 16 pp.	$0 10
DRAINAGE TABLES. By Geo. H. Johnson, C.E. Paper, 5 x 7½ inches, 15 pp.	25
ELEMENTS OF WATER-SUPPLY ENGINEERING. By E. Sherman Gould, M. Am. Soc. C. E. (In preparation.)	
ENGINEERING CONTRACTS AND SPECIFICATIONS. By J. B. Johnson, M. I. C. E., M. Am. Soc. C. E., M. Am. Soc. M. E., Professor of Civil Engineering, Washington University. Cloth, 6x9 inches, 452 pp. with complete index. New edition.	4 00
MANUAL OF AMERICAN WATER WORKS. By M. N. Baker, Associate Editor "Engineering News." Cloth, 6 x 9 inches, 700 pp.	3 00
ROAD MAKING AND MAINTENANCE. By Clemens Herschel, M. Am. Soc. C. E., and Edward P. North, M. Am. Soc. C. E. Paper, 6 x 9 inches, 156 pp.	50
SEWAGE DISPOSAL IN THE UNITED STATES. By Geo. W. Rafter, M. Am. Soc. C. E., and M. N. Baker, Associate Editor "Engineering News." Cloth, 7 x 10 inches, 598 pp.; 7 plates and 116 illustrations in the text.	6 00
TABLES AND DIAGRAMS FOR ESTIMATING SEWER WORK. By S. M. Swaab, C.E. Paper, oblong, 4½ x 7½ inches, 20 pp., 16 plates.	50
TABLES FOR ESTIMATING THE COST OF LAYING CAST-IRON WATER PIPE. By Edmund B. Weston. M. Am. Soc. C. E., M. Inst. C. E. Pamphlet, 6 x 9 inches, 12 pp.	25

Any Standard Books on Engineering and Allied Subjects may be procured from our New York and Chicago offices.

A complete list of our own books or a sample copy of "Engineering News" will be mailed on application.

The Engineering News Publishing Co.,

Chicago Branch: 1636 Monadnock Block.   Publication Office: 220 Broadway, New York.

SHORT-TITLE CATALOGUE

OF THE

PUBLICATIONS

OF

JOHN WILEY & SONS,

NEW YORK.

LONDON: CHAPMAN & HALL, LIMITED.

ARRANGED UNDER SUBJECTS.

Descriptive circulars sent on application.
Books marked with an asterisk are sold at *net* prices only.
All books are bound in cloth unless otherwise stated.

AGRICULTURE.

CATTLE FEEDING—DAIRY PRACTICE—DISEASES OF ANIMALS—GARDENING, ETC.

Armsby's Manual of Cattle Feeding................12mo,	$1	75
Downing's Fruit and Fruit Trees.......................8vo,	5	00
Grotenfelt's The Principles of Modern Dairy Practice. (Woll.) 12mo,	2	00
Kemp's Landscape Gardening.........................12mo,	2	50
Loudon's Gardening for Ladies. (Downing.)........12mo,	1	50
Maynard's Landscape Gardening.....................12mo,	1	50
Steel's Treatise on the Diseases of the Dog...........8vo,	3	50
" Treatise on the Diseases of the Ox..............8vo,	6	00
Stockbridge's Rocks and Soils........................8vo,	2	50
Woll's Handbook for Farmers and Dairymen........12mo,	1	50

ARCHITECTURE.

BUILDING—CARPENTRY—STAIRS—VENTILATION—LAW, ETC.

Berg's Buildings and Structures of American Railroads.....4to,	7	50
Birkmire's American Theatres—Planning and Construction.8vo,	3	00
" Architectural Iron and Steel.....................8vo,	3	50
" Compound Riveted Girders.....................8vo,	2	00
" Skeleton Construction in Buildings............8vo,	3	00

Birkmire's Planning and Construction of High Office Buildings. 8vo,	$3	50
Carpenter's Heating and Ventilating of Buildings..........8vo,	3	00
Freitag's Architectural Engineering.................8vo,	2	50
Gerhard's Sanitary House Inspection..................16mo,	1	00
" Theatre Fires and Panics................12mo,	1	50
Hatfield's American House Carpenter..................8vo,	5	00
Holly's Carpenter and Joiner..........................18mo,		75
Kidder's Architect and Builder's Pocket-book...16mo, morocco,	4	00
Merrill's Stones for Building and Decoration.............8vo,	5	00
Monckton's Stair Building—Wood, Iron, and Stone........4to,	4	00
Wait's Engineering and Architectural Jurisprudence.......8vo,	6	00
Sheep,	6	50
Worcester's Small Hospitals—Establishment and Maintenance, including Atkinson's Suggestions for Hospital Architecture...12mo,	1	25
World's Columbian Exposition of 1893.............Large 4to,	2	50

ARMY, NAVY, Etc.

MILITARY ENGINEERING—ORDNANCE—LAW, ETC.

Bourne's Screw Propellers..............................4to,	5	00
* Bruff's Ordnance and Gunnery......................8vo,	6	00
Chase's Screw Propellers............................8vo,	3	00
Cooke's Naval Ordnance 8vo,	12	50
Cronkhite's Gunnery for Non-com. Officers......32mo, morocco,	2	00
* Davis's Treatise on Military Law.......................8vo,	7	00
Sheep,	7	50
* " Elements of Law............................8vo,	2	50
De Brack's Cavalry Outpost Duties. (Carr.)....32mo, morocco,	2	00
Dietz's Soldier's First Aid.....................16mo, morocco,	1	25
* Dredge's Modern French Artillery....Large 4to, half morocco,	15	00
" Record of the Transportation Exhibits Building, World's Columbian Exposition of 1893..4to, half morocco,	10	00
Durand's Resistance and Propulsion of Ships..............8vo,	5	00
Dyer's Light Artillery................................12mo,	3	00
Hoff's Naval Tactics...............................8vo,	1	50
* Ingalls's Ballistic Tables..............................8vo,	1	50
" Handbook of Problems in Direct Fire............8vo,	4	00

2

Mahan's Permanent Fortifications. (Mercur.).8vo, half morocco,	$7 50
Mercur's Attack of Fortified Places....................12mo,	2 00
" Elements of the Art of War......................8vo,	4 00
Metcalfe's Ordnance and Gunnery...........12mo, with Atlas,	5 00
Murray's A Manual for Courts-Martial....... 16mo, morocco,	1 50
" Infantry Drill Regulations adapted to the Springfield Rifle, Caliber .45......................32mo, paper,	10
* Phelps's Practical Marine Surveying....................8vo,	2 50
Powell's Army Officer's Examiner....................12mo,	4 00
Sharpe's Subsisting Armies................ 32mo, morocco,	1 50
Very's Navies of the World...............8vo, half morocco,	3 50
Wheeler's Siege Operations...........................8vo,	2 00
Winthrop's Abridgment of Military Law...............12mo,	2 50
Woodhull's Notes on Military Hygiene................16mo,	1 50
Young's Simple Elements of Navigation ......16mo, morocco,	2 00
" " " " " first edition........	1 00

ASSAYING.

SMELTING—ORE DRESSING—ALLOYS, ETC.

Fletcher's Quant. Assaying with the Blowpipe..16mo, morocco,	1 50
Furman's Practical Assaying...........................8vo,	3 00
Kunhardt's Ore Dressing. 8vo,	1 50
O'Driscoll's Treatment of Gold Ores...................8vo,	2 00
Ricketts and Miller's Notes on Assaying................8vo,	3 00
Thurston's Alloys, Brasses, and Bronzes 8vo,	2 50
Wilson's Cyanide Processes 12mo,	1 50
" The Chlorination Process....................12mo,	1 50

ASTRONOMY.

PRACTICAL, THEORETICAL, AND DESCRIPTIVE.

Craig's Azimuth 4to,	3 50
Doolittle's Practical Astronomy.......................8vo,	4 00
Gore's Elements of Geodesy........................ 8vo,	2 50
Hayford's Text-book of Geodetic Astronomy............8vo,	3 00
* Michie and Harlow's Practical Astronomy..............8vo,	3 00
* White's Theoretical and Descriptive Astronomy........12mo,	2 00

BOTANY.

GARDENING FOR LADIES, ETC.

Baldwin's Orchids of New England..................Small 8vo,	$1 50
Loudon's Gardening for Ladies. (Downing.)............12mo,	1 50
Thomé's Structural Botany..............................16mo,	2 25
Westermaier's General Botany. (Schneider.)..............8vo,	2 00

BRIDGES, ROOFS, Etc.

CANTILEVER—DRAW—HIGHWAY—SUSPENSION.
(*See also* ENGINEERING, p. 7.)

Boller's Highway Bridges.................................8vo,	2 00
* " The Thames River Bridge.................4to, paper,	5 00
Burr's Stresses in Bridges..................................8vo,	3 50
Crehore's Mechanics of the Girder........................8vo,	5 00
Dredge's Thames Bridges............... ....7 parts, per part,	1 25
Du Bois's Stresses in Framed Structures.............Small 4to,	10 00
Foster's Wooden Trestle Bridges..........................4to,	5 00
Greene's Arches in Wood, etc............................8vo,	2 50
" Bridge Trusses.................................8vo,	2 50
" Roof Trusses..................................8vo,	1 25
Howe's Treatise on Arches 8vo,	4 00
Johnson's Modern Framed Structures................Small 4to,	10 00
Merriman & Jacoby's Text-book of Roofs and Bridges. Part I., Stresses......................................8vo,	2 50
Merriman & Jacoby's Text-book of Roofs and Bridges. Part II., Graphic Statics 8vo,	2 50
Merriman & Jacoby's Text-book of Roofs and Bridges. Part III., Bridge Design............................8vo,	2 50
Merriman & Jacoby's Text-book of Roofs and Bridges. Part IV., Continuous, Draw, Cantilever, Suspension, and Arched Bridges....................................8vo,	2 50
* Morison's The Memphis Bridge..................Oblong 4to,	10 00
Waddell's Iron Highway Bridges..... ....8vo,	4 00
" De Pontibus (a Pocket-book for Bridge Engineers). 16mo, morocco,	3 00
Wood's Construction of Bridges and Roofs................8vo,	2 00
Wright's Designing of Draw Spans. Parts I. and II..8vo, each	2 50
" " " " " Complete...........8vo,	3 50

CHEMISTRY.

QUALITATIVE—QUANTITATIVE—ORGANIC—INORGANIC, ETC.

Adriance's Laboratory Calculations..................12mo,	$1 25
Allen's Tables for Iron Analysis........................8vo,	3 00
Austen's Notes for Chemical Students..................12mo,	1 50
Bolton's Student's Guide in Quantitative Analysis........8vo,	1 50
Classen's Analysis by Electrolysis. (Herrick and Boltwood.).8vo,	3 00
Crafts's Qualitative Analysis. (Schaeffer.)..............12mo,	1 50
Drechsel's Chemical Reactions. (Merrill.)..............12mo,	1 25
Fresenius's Quantitative Chemical Analysis. (Allen.).......8vo,	6 00
" Qualitative " " (Johnson.).....8vo,	3 00
" " " " (Wells.) Trans.	
16th German Edition............................8vo,	5 00
Fuertes's Water and Public Health.....................12mo,	1 50
Gill's Gas and Fuel Analysis..........................12mo,	1 25
Hammarsten's Physiological Chemistry. (Mandel.).......8vo,	4 00
Helm's Principles of Mathematical Chemistry. (Morgan).12mo,	1 50
Kolbe's Inorganic Chemistry..........................12mo,	1 50
Ladd's Quantitative Chemical Analysis.................12mo,	1 00
Landauer's Spectrum Analysis. (Tingle.)................8vo,	3 00
Löb's Electrolysis and Electrosynthesis of Organic Compounds.	
(Lorenz.)......................................12mo,	1 00
Mandel's Bio-chemical Laboratory.....................12mo,	1 50
Mason's Water-supply...................................8vo,	5 00
" Examination of Water........................12mo,	1 25
Meyer's Organic Analysis. (Tingle.) (*In the press.*)	
Miller's Chemical Physics..............................8vo,	2 00
Mixter's Elementary Text-book of Chemistry............12mo,	1 50
Morgan's The Theory of Solutions and its Results.......12mo,	1 00
" Elements of Physical Chemistry..............12mo,	2 00
Nichols's Water-supply (Chemical and Sanitary)..........8vo,	2 50
O'Brine's Laboratory Guide to Chemical Analysis.........8vo,	2 00
Perkins's Qualitative Analysis..........................12mo,	1 00
Pinner's Organic Chemistry. (Austen.)................12mo,	1 50
Poole's Calorific Power of Fuels.........................8vo,	3 00
Ricketts and Russell's Notes on Inorganic Chemistry (Non-metallic)....................Oblong 8vo, morocco,	75
Ruddiman's Incompatibilities in Prescriptions.............8vo,	2 00

Schimpf's Volumetric Analysis............12mo,	$2 50
Spencer's Sugar Manufacturer's Handbook.....16mo, morocco,	2 00
" Handbook for Chemists of Beet Sugar Houses. 16mo, morocco,	3 00
Stockbridge's Rocks and Soils............8vo,	2 50
Tillman's Descriptive General Chemistry. (*In the press.*)	
Van Deventer's Physical Chemistry for Beginners. (Boltwood.) 12mo,	1 50
Wells's Inorganic Qualitative Analysis............12mo,	1 50
" Laboratory Guide in Qualitative Chemical Analysis. 8vo,	1 50
Whipple's Microscopy of Drinking-water...........8vo,	3 50
Wiechmann's Chemical Lecture Notes............12mo,	3 00
" Sugar Analysis............Small 8vo,	2 50
Wulling's Inorganic Phar. and Med. Chemistry.........12mo,	2 00

DRAWING.

ELEMENTARY—GEOMETRICAL—MECHANICAL—TOPOGRAPHICAL.

Hill's Shades and Shadows and Perspective............8vo,	2 00
MacCord's Descriptive Geometry............8vo,	3 00
" Kinematics............8vo,	5 00
" Mechanical Drawing............8vo,	4 00
Mahan's Industrial Drawing. (Thompson.)......2 vols., 8vo,	3 50
Reed's Topographical Drawing. (H. A.)............4to,	5 00
Reid's A Course in Mechanical Drawing............8vo,	2 00
" Mechanical Drawing and Elementary Machine Design. 8vo. (*In the press.*)	
Smith's Topographical Drawing. (Macmillan.)............8vo,	2 50
Warren's Descriptive Geometry............2 vols., 8vo,	3 50
" Drafting Instruments............12mo,	1 25
" Free-hand Drawing............12mo,	1 00
" Linear Perspective............12mo,	1 00
" Machine Construction............2 vols., 8vo,	7 50
" Plane Problems............12mo,	1 25
" Primary Geometry............12mo,	75
" Problems and Theorems............8vo,	2 50
" Projection Drawing............12mo,	1 50

Warren's Shades and Shadows..........................8vo,	$3	00
" Stereotomy—Stone-cutting...................8vo,	2	50
Whelpley's Letter Engraving.......................12mo,	2	00

ELECTRICITY AND MAGNETISM.

ILLUMINATION—BATTERIES—PHYSICS—RAILWAYS.

Anthony and Brackett's Text-book of Physics. (Magie.) Small 8vo,	3	00
Anthony's Theory of Electrical Measurements..........12mo,	1	00
Barker's Deep-sea Soundings..........................8vo,	2	00
Benjamin's Voltaic Cell...............................8vo,	3	00
" History of Electricity..........................8vo,	3	00
Classen's Analysis by Electrolysis. (Herrick and Boltwood.) 8vo,	3	00
Cosmic Law of Thermal Repulsion....................12mo,		75
Crehore and Squier's Experiments with a New Polarizing Photo-Chronograph...8vo,	*3	00
Dawson's Electric Railways and Tramways. Small, 4to, half morocco,	12	50
*Dredge's Electric Illuminations....2 vols., 4to, half morocco,	25	00
" " " Vol. II..................4to,	7	50
Gilbert's De magnete. (Mottelay.).....................8vo,	2	50
Holman's Precision of Measurements. 8vo,	2	00
" Telescope-mirror-scale Method..........Large 8vo,		75
Löb's Electrolysis and Electrosynthesis of Organic Compounds. (Lorenz.)..12mo,	1	00
*Michie's Wave Motion Relating to Sound and Light......8vo,	4	00
Morgan's The Theory of Solutions and its Results........12mo,	1	00
Niaudet's Electric Batteries. (Fishback.)..............12mo,	2	50
Pratt and Alden's Street-railway Road-beds.............8vo,	2	00
Reagan's Steam and Electric Locomotives..............12mo,	2	00
Thurston's Stationary Steam Engines for Electric Lighting Purposes..8vo,	2	50
*Tillman's Heat......................................8vo,	1	50

ENGINEERING.

CIVIL—MECHANICAL—SANITARY, ETC.

(*See also* BRIDGES, p. 4; HYDRAULICS, p. 9; MATERIALS OF ENGINEERING, p. 10; MECHANICS AND MACHINERY, p. 12; STEAM ENGINES AND BOILERS, p. 14.)

Baker's Masonry Construction..........................8vo,	$5 00
" Surveying Instruments.......................12mo,	3 00
Black's U. S. Public Works......................Oblong 4to,	5 00
Brooks's Street-railway Location...............16mo, morocco,	1 50
Butts's Civil Engineers' Field Book............16mo, morocco,	2 50
Byrne's Highway Construction..............................8vo,	5 00
" Inspection of Materials and Workmanship.......16mo,	3 00
Carpenter's Experimental Engineering 8vo,	6 00
Church's Mechanics of Engineering—Solids and Fluids....8vo,	6 00
" Notes and Examples in Mechanics...............8vo,	2 00
Crandall's Earthwork Tables.......................8vo,	1 50
" The Transition Curve...............16mo, morocco,	1 50
* Dredge's Penn. Railroad Construction, etc. Large 4to,	
half morocco,	20 00
* Drinker's Tunnelling.....................4to, half morocco,	25 00
Eissler's Explosives—Nitroglycerine and Dynamite........8vo,	4 00
Folwell's Sewerage..8vo,	3 00
Fowler's Coffer-dam Process for Piers....................8vo,	2 50
Gerhard's Sanitary House Inspection....................12mo,	1 00
Godwin's Railroad Engineer's Field-book......16mo, morocco,	2 50
Gore's Elements of Geodesy........ ....8vo,	2 50
Howard's Transition Curve Field-book.........16mo, morocco,	1 50
Howe's Retaining Walls (New Edition.)................12mo,	1 25
Hudson's Excavation Tables. Vol. II................. 8vo,	1 00
Hutton's Mechanical Engineering of Power Plants........8vo,	5 00
Johnson's Materials of Construction...............Large 8vo,	6 00
" Stadia Reduction Diagram..Sheet, 22½ × 28½ inches,	50
" Theory and Practice of Surveying........Small 8vo,	4 00
Kent's Mechanical Engineer's Pocket-book.....16mo, morocco,	5 00
Kiersted's Sewage Disposal..... ....12mo,	1 25
Mahan's Civil Engineering. (Wood.)....................8vo,	5 00
Merriman and Brook's Handbook for Surveyors....16mo, mor.,	2 00
Merriman's Geodetic Surveying..........................8vo,	2 00
" Retaining Walls and Masonry Dams..........8vo,	2 00
" Sanitary Engineering......................8vo,	2 00
Nagle's Manual for Railroad Engineers........16mo, morocco,	3 00
Ogden's Sewer Design. (*In the press.*)	
Patton's Civil Engineering.................8vo, half morocco,	7 50

Patton's Foundations............8vo,	$5 00
Pratt and Alden's Street-railway Road-beds............8vo,	2 00
Rockwell's Roads and Pavements in France........12mo,	1 25
Searles's Field Engineering ....16mo, morocco,	3 00
" Railroad Spiral........................16mo, morocco,	1 50
Siebert and Biggin's Modern Stone Cutting and Masonry...8vo,	1 50
Smart's Engineering Laboratory Practice.............12mo,	2 50
Smith's Wire Manufacture and Uses..............Small 4to,	3 00
Spalding's Roads and Pavements..............12mo,	2 00
" Hydraulic Cement.......................12mo,	2 00
Taylor's Prismoidal Formulas and Earthwork............8vo,	1 50
Thurston's Materials of Construction 8vo,	5 00
* Trautwine's Civil Engineer's Pocket-book....16mo, morocco,	5 00
* " Cross-section........................Sheet,	25
* " Excavations and Embankments...........8vo,	2 00
* " Laying Out Curves...........12mo, morocco,	2 50
Waddell's De Pontibus (A Pocket-book for Bridge Engineers). 16mo, morocco,	3 00
Wait's Engineering and Architectural Jurisprudence.......8vo,	6 00
Sheep,	6 50
" Law of Field Operation in Engineering, etc........8vo.	
Warren's Stereotomy—Stone-cutting.....................8vo,	2 50
*Webb's Engineering Instruments............16mo, morocco,	50
" " " New Edition.............	1 25
Wegmann's Construction of Masonry Dams.............4to,	5 00
Wellington's Location of Railways..............Small 8vo,	5 00
Wheeler's Civil Engineering..........................8vo,	4 00
Wolff's Windmill as a Prime Mover....................8vo,	3 00

HYDRAULICS.

WATER-WHEELS—WINDMILLS—SERVICE PIPE—DRAINAGE, ETC.

(*See also* ENGINEERING, p. 7.)

Bazin's Experiments upon the Contraction of the Liquid Vein. (Trautwine.).........................8vo,	2 00
Bovey's Treatise on Hydraulics.......................8vo,	4 00
Coffin's Graphical Solution of Hydraulic Problems........12mo,	2 50
Ferrel's Treatise on the Winds, Cyclones, and Tornadoes...8vo,	4 00
Fuertes's Water and Public Health.....................12mo,	1 50
Ganguillet & Kutter's Flow of Water. (Hering & Trautwine.) 8vo,	4 00
Hazen's Filtration of Public Water Supply...............8vo,	2 00
Herschel's 115 Experiments...........................8vo,	2 00

Kiersted's Sewage Disposal..............................12mo,	$1 25
Mason's Water Supply......................................8vo,	5 00
" Examination of Water. 12mo,	1 25
Merriman's Treatise on Hydraulics......................8vo,	4 00
Nichols's Water Supply (Chemical and Sanitary)..........8vo,	2 50
Wegmann's Water Supply of the City of New York........4to,	10 00
Weisbach's Hydraulics. (Du Bois.)........................8vo,	5 00
Whipple's Microscopy of Drinking Water.................8vo,	3 50
Wilson's Irrigation Engineering............................8vo,	4 00
" Hydraulic and Placer Mining...................12mo,	2 00
Wolff's Windmill as a Prime Mover........................8vo,	3 00
Wood's Theory of Turbines...................................8vo,	2 50

MANUFACTURES.

BOILERS—EXPLOSIVES—IRON—STEEL—SUGAR—WOOLLENS, ETC.

Allen's Tables for Iron Analysis............................8vo,	3 00
Beaumont's Woollen and Worsted Manufacture.........12mo,	1 50
Bolland's Encyclopædia of Founding Terms............12mo,	3 00
" The Iron Founder..12mo,	2 50
" " " " Supplement........................12mo,	2 50
Bouvier's Handbook on Oil Painting......................12mo,	2 00
Eissler's Explosives, Nitroglycerine and Dynamite........8vo,	4 00
Fodr's Boiler Making for Boiler Makers.................18mo,	1 00
Metcalfe's Cost of Manufactures............................8vo,	5 00
Metcalf's Steel—A Manual for Steel Users...............12mo,	2 00
*Reisig's Guide to Piece Dyeing............................8vo,	25 00
Spencer's Sugar Manufacturer's Handbook....16mo, morocco,	2 00
" Handbook for Chemists of Beet Sugar Houses. 16mo, morocco,	3 00
Thurston's Manual of Steam Boilers........................ 8vo,	5 00
Walke's Lectures on Explosives..............................8vo,	4 00
West's American Foundry Practice.......................12mo,	2 50
" Moulder's Text-book 12mo,	2 50
Wiechmann's Sugar Analysis...................... Small 8vo,	2 50
Woodbury's Fire Protection of Mills......................8vo,	2 50

MATERIALS OF ENGINEERING.

STRENGTH—ELASTICITY—RESISTANCE, ETC.

(*See also* ENGINEERING, p. 7.)

Baker's Masonry Construction..............................8vo,	5 00
Beardslee and Kent's Strength of Wrought Iron...........8vo,	1 50
Bovey's Strength of Materials................................8vo,	7 50
Burr's Elasticity and Resistance of Materials...............8vo,	5 00
Byrne's Highway Construction...............................8vo,	5 00

Church's Mechanics of Engineering—Solids and Fluids 8vo,	$6	00
Du Bois's Stresses in Framed Structures............Small 4to,	10	00
Johnson's Materials of Construction......................8vo,	6	00
Lanza's Applied Mechanics................................8vo,	7	50
Martens's Materials. (Henning.)..........8vo. (*In the press.*)		
Merrill's Stones for Building and Decoration..............8vo,	5	00
Merriman's Mechanics of Materials........................8vo,	4	00
" Strength of Materials........................12mo,	1	00
Patton's Treatise on Foundations..........................8vo,	5	00
Rockwell's Roads and Pavements in France.............12mo,	1	25
Spalding's Roads and Pavements........................12mo,	2	00
Thurston's Materials of Construction......................8vo,	5	00
" Materials of Engineering...............3 vols., 8vo,	8	00
Vol. I., Non-metallic 8vo,	2	00
Vol. II., Iron and Steel........................8vo,	3	50
Vol. III., Alloys, Brasses, and Bronzes.............8vo,	2	50
Wood's Resistance of Materials...........................8vo,	2	00

MATHEMATICS.

CALCULUS—GEOMETRY—TRIGONOMETRY, ETC.

Baker's Elliptic Functions..................................8vo,	1	50
Ballard's Pyramid Problem................................8vo,	1	50
Barnard's Pyramid Problem................................8vo,	1	50
*Bass's Differential Calculus............................12mo,	4	00
Briggs's Plane Analytical Geometry.....................12mo,	1	00
Chapman's Theory of Equations........................12mo,	1	50
Compton's Logarithmic Computations..................12mo,	1	50
Davis's Introduction to the Logic of Algebra..............8vo,	1	50
Halsted's Elements of Geometry..........................8vo,	1	75
" Synthetic Geometry..........................8vo,	1	50
Johnson's Curve Tracing................................12mo,	1	00
" Differential Equations—Ordinary and Partial. Small 8vo,	3	50
" Integral Calculus............................12mo,	1	50
" " " Unabridged. Small 8vo. (*In the press.*)		
" Least Squares..............................12mo,	1	50
*Ludlow's Logarithmic and Other Tables. (Bass.).......8vo,	2	00
* " Trigonometry with Tables. (Bass.)............8vo,	3	00
*Mahan's Descriptive Geometry (Stone Cutting) 8vo,	1	50
Merriman and Woodward's Higher Mathematics........8vo,	5	00
Merriman's Method of Least Squares....................8vo,	2	00
Parker's Quadrature of the Circle........................8vo,	2	50
Rice and Johnson's Differential and Integral Calculus, 2 vols. in 1, small 8vo,	2	50

Rice and Johnson's Differential Calculus............Small 8vo,		$3 00
" Abridgment of Differential Calculus. Small 8vo,		1 50
Totten's Metrology...8vo,		2 50
Warren's Descriptive Geometry..................2 vols., 8vo,		3 50
" Drafting Instruments.......................12mo,		1 25
" Free-hand Drawing.........................12mo,		1 00
" Higher Linear Perspective....................8vo,		3 50
" Linear Perspective.........................12mo,		1 00
" Primary Geometry..........................12mo,		75
" Plane Problems............................12mo,		1 25
" Problems and Theorems......................8vo,		2 50
" Projection Drawing........................12mo,		1 50
Wood's Co-ordinate Geometry..............................8vo,		2 00
" Trigonometry..............................12mo,		1 00
Woolf's Descriptive Geometry.......................Large 8vo,		3 00

MECHANICS—MACHINERY.

TEXT-BOOKS AND PRACTICAL WORKS.

(*See also* ENGINEERING, p. 7.)

Baldwin's Steam Heating for Buildings.................12mo,		2 50
Benjamin's Wrinkles and Recipes.......................12mo,		2 00
Chordal's Letters to Mechanics........................12mo,		2 00
Church's Mechanics of Engineering......................8vo,		6 00
" Notes and Examples in Mechanics..............8vo,		2 00
Crehore's Mechanics of the Girder.......................8vo,		5 00
Cromwell's Belts and Pulleys..........................12mo,		1 50
" Toothed Gearing...........................12mo,		1 50
Compton's First Lessons in Metal Working..............12mo,		1 50
Compton and De Groodt's Speed Lathe..................12mo,		1 50
Dana's Elementary Mechanics..........................12mo,		1 50
Dingey's Machinery Pattern Making....................12mo,		2 00
Dredge's Trans. Exhibits Building, World Exposition. Large 4to, half morocco,		10 00
Du Bois's Mechanics. Vol. I., Kinematics..............8vo,		3 50
" " Vol. II., Statics..................8vo,		4 00
" " Vol. III., Kinetics.................8vo,		3 50
Fitzgerald's Boston Machinist.........................18mo,		1 00
Flather's Dynamometers...............................12mo,		2 00
" Rope Driving..............................12mo,		2 00
Hall's Car Lubrication................................12mo,		1 00
Holly's Saw Filing....................................18mo,		75
Johnson's Theoretical Mechanics. An Elementary Treatise (*In the press.*)		
Jones's Machine Design. Part I., Kinematics............8vo,		1 50

Jones's Machine Design. Part II., Strength and Proportion of Machine Parts............8vo,	$3	00
Lanza's Applied Mechanics............8vo,	7	50
MacCord's Kinematics............8vo,	5	00
Merriman's Mechanics of Materials............8vo,	4	00
Metcalfe's Cost of Manufactures............8vo,	5	00
*Michie's Analytical Mechanics............8vo,	4	00
Richards's Compressed Air............12mo,	1	50
Robinson's Principles of Mechanism............8vo,	3	00
Smith's Press-working of Metals............8vo,	3	00
Thurston's Friction and Lost Work............8vo,	3	00
" The Animal as a Machine............12mo,	1	00
Warren's Machine Construction............2 vols., 8vo,	7	50
Weisbach's Hydraulics and Hydraulic Motors. (Du Bois.)..8vo,	5	00
" Mechanics of Engineering. Vol. III., Part I., Sec. I. (Klein.)............8vo,	5	00
Weisbach's Mechanics of Engineering. Vol. III., Part I., Sec. II. (Klein.)............8vo,	5	00
Weisbach's Steam Engines. (Du Bois.)............8vo,	5	00
Wood's Analytical Mechanics............8vo,	3	00
" Elementary Mechanics............12mo,	1	25
" " " Supplement and Key.....12mo,	1	25

METALLURGY.

IRON—GOLD—SILVER—ALLOYS, ETC.

Allen's Tables for Iron Analysis............8vo,	3	00
Egleston's Gold and Mercury............Large 8vo,	7	50
" Metallurgy of Silver............Large 8vo,	7	50
* Kerl's Metallurgy—Copper and Iron............8vo,	15	00
* " " Steel, Fuel, etc............8vo,	15	00
Kunhardt's Ore Dressing in Europe............8vo,	1	50
Metcalf's Steel—A Manual for Steel Users............12mo,	2	00
O'Driscoll's Treatment of Gold Ores............8vo,	2	00
Thurston's Iron and Steel............8vo,	3	50
" Alloys............8vo,	2	50
Wilson's Cyanide Processes............12mo,	1	50

MINERALOGY AND MINING.

MINE ACCIDENTS—VENTILATION—ORE DRESSING, ETC.

Barringer's Minerals of Commercial Value....Oblong morocco,	2	50
Beard's Ventilation of Mines............12mo,	2	50
Boyd's Resources of South Western Virginia............8vo,	3	00
" Map of South Western Virginia.....Pocket-book form,	2	00

Brush and Penfield's Determinative Mineralogy. New Ed. 8vo,	$4 00
Chester's Catalogue of Minerals....................8vo,	1 25
" " " " ....Paper,	50
" Dictionary of the Names of Minerals...........8vo,	3 00
Dana's American Localities of Minerals............Large 8vo,	1 00
" Descriptive Mineralogy. (E. S.)....Large half morocco,	12 50
" Mineralogy and Petrography. (J. D.)..........12mo,	2 00
" Minerals and How to Study Them. (E. S.)......12mo,	1 50
" Text-book of Mineralogy. (E. S.)...New Edition. 8vo,	4 00
* Drinker's Tunnelling, Explosives, Compounds, and Rock Drills. 4to, half morocco,	25 00
Egleston's Catalogue of Minerals and Synonyms...........8vo,	2 50
Eissler's Explosives—Nitroglycerine and Dynamite........8vo,	4 00
Hussak's Rock-forming Minerals. (Smith.).........Small 8vo,	2 00
Ihlseng's Manual of Mining. 8vo,	4 00
Kunhardt's Ore Dressing in Europe.....................8vo,	1 50
O'Driscoll's Treatment of Gold Ores......................8vo,	2 00
* **Penfield's Record** of Mineral Tests............Paper, 8vo,	50
Rosenbusch's Microscopical Physiography of Minerals and Rocks. (Iddings.)................................8vo,	5 00
Sawyer's Accidents in Mines......................Large 8vo,	7 00
Stockbridge's Rocks and Soils...........................8vo,	2 50
Walke's Lectures on Explosives.........................8vo,	4 00
Williams's Lithology 8vo,	3 00
Wilson's Mine Ventilation..............................12mo,	1 25
" Hydraulic and Placer Mining..... ....12mo,	2 50

STEAM AND ELECTRICAL ENGINES, BOILERS, Etc.

STATIONARY—MARINE—LOCOMOTIVE—GAS ENGINES, ETC.

(*See also* ENGINEERING, p. 7.)

Baldwin's Steam Heating for Buildings................12mo,	2 50
Clerk's Gas Engine.........................Small 8vo,	4 00
Ford's Boiler Making for Boiler Makers................18mo,	1 00
Hemenway's Indicator Practice........................12mo,	2 00
Hoadley's Warm-blast Furnace..........................8vo,	1 50
Kneass's Practice and Theory of the Injector 8vo,	1 50
MacCord's Slide Valve................................8vo,	2 00
Meyer's Modern Locomotive Construction................4to,	10 00
Peabody and Miller's Steam-boilers.....................8vo,	4 00
Peabody's Tables of Saturated Steam...................8vo,	1 00
" Thermodynamics of the Steam Engine......... 8vo,	5 00
" Valve Gears for the Steam Engine............8vo,	2 50
Pray's Twenty Years with the Indicator...........Large 8vo,	2 50
Pupin and Osterberg's Thermodynamics...............12mo,	1 25

Reagan's Steam and E'ectric Locomotives................12mo,	$2	00
Röntgen's Thermodynamics. (Du Bois.)..............8vo,	5	00
Sinclair's Locomotive Running...................12mo,	2	00
Snow's Steam-boiler Practice........8vo. (*In the press*.)		
Thurston's Boiler Explosions..................12mo,	1	50
" Engine and Boiler Trials...................8vo,	5	00
" Manual of the Steam Engine. Part I., Structure and Theory............................8vo,	6	00
" Manual of the Steam Engine. Part II., Design, Construction, and Operation..............8vo,	6	00
2 parts,	10	00
Thurston's Philosophy of the Steam Engine............12mo,		75
" Reflection on the Motive Power of Heat. (Carnot.) 12mo,	1	50
" Stationary Steam Engines....................8vo,	2	50
" Steam-boiler Construction and Operation......8vo,	5	00
Spangler's Valve Gears........................8vo,	2	50
Weisbach's Steam Engine. (Du Bois.).............8vo,	5	00
Whitham's Constructive Steam Engineering..............8vo,	6	00
" Steam-engine Design................8vo,	5	00
Wilson's Steam Boilers. (Flather.)...................12mo,	2	50
Wood's Thermodynamics, Heat Motors, etc.............8vo,	4	00

TABLES, WEIGHTS, AND MEASURES.

FOR ACTUARIES, CHEMISTS, ENGINEERS, MECHANICS—METRIC TABLES, ETC.

Adriance's Laboratory Calculations..................12mo,	1	25
Allen's Tables for Iron Analysis......................8vo,	3	00
Bixby's Graphical Computing Tables...........Sheet,		25
Compton's Logarithms.......................12mo,	1	50
Crandall's Railway and **Earthwork** Tables...........8vo,	1	50
Egleston's Weights and Measures................18mo,		75
Fisher's Table of Cubic Yards.............Cardboard,		25
Hudson's Excavation Tables. Vol. II..............8vo,	1	00
Johnson's Stadia and Earthwork Tables............8vo,	1	25
Ludlow's Logarithmic and Other Tables. (Bass.).......12mo,	2	00
Totten's Metrology..........................8vo,	2	50

VENTILATION.

STEAM HEATING—HOUSE INSPECTION—MINE **VENTILATION.**

Baldwin's Steam Heating........................12mo,	2	50
Beard's Ventilation of Mines....................12mo,	2	50
Carpenter's Heating and Ventilating of Buildings........8vo,	3	00
Gerhard's Sanitary House Inspection................12mo,	1	00
Reid's **Ventilation of** American Dwellings..........12mo,	1	50
Wilson**'s Mine Ventilation**.......................12mo,	1	25

MISCELLANEOUS PUBLICATIONS.

Alcott's Gems, Sentiment, Language............Gilt edges,	$5 00
Bailey's The New Tale of a Tub........................8vo,	75
Ballard's Solution of the Pyramid Problem..............8vo,	1 50
Barnard's The Metrological System of the Great Pyramid..8vo,	1 50
Davis's Elements of Law..............................8vo,	2 00
Emmon's Geological Guide-book of the Rocky Mountains..8vo,	1 50
Ferrel's Treatise on the Winds........................8vo,	4 00
Haines's Addresses Delivered before the Am. Ry. Assn...12mo,	2 50
Mott's The Fallacy of the Present Theory of Sound..Sq. 16mo,	1 00
Perkins's Cornell University....................Oblong 4to,	1 50
Ricketts's History of Rensselaer Polytechnic Institute.....8vo,	3 00
Rotherham's The New Testament Critically Emphasized. 12mo,	1 50
" The Emphasized New Test. A new translation. Large 8vo,	2 00
Totten's An Important Question in Metrology............8vo,	2 50
Whitehouse's Lake Mœris..............................Paper,	25
* Wiley's Yosemite, Alaska, and Yellowstone............4to,	3 00

HEBREW AND CHALDEE TEXT-BOOKS.

For Schools and Theological Seminaries.

Gesenius's Hebrew and Chaldee Lexicon to Old Testament. (Tregelles,)...................Small 4to, half morocco,	5 00
Green's Elementary Hebrew Grammar..................12mo,	1 25
" Grammar of the Hebrew Language (New Edition).8vo,	3 00
" Hebrew Chrestomathy.........................8vo,	2 00
Letteris's Hebrew Bible (Massoretic Notes in English). 8vo, arabesque,	2 25

MEDICAL.

Bull's Maternal Management in Health and Disease.......12mo,	1 00
Hammarsten's Physiological Chemistry. (Mandel.)........8vo,	4 00
Mott's Composition, Digestibility, and Nutritive Value of Food. Large mounted chart,	1 25
Ruddiman's Incompatibilities in Prescriptions............8vo,	2 00
Steel's Treatise on the Diseases of the Ox...............8vo,	6 00
" Treatise on the Diseases of the Dog...............8vo,	3 50
Woodhull's Military Hygiene..........................16mo,	1 50
Worcester's Small Hospitals—Establishment and Maintenance, including Atkinson's Suggestions for Hospital Architecture...12mo,	1 25

www.ingramcontent.com/pod-product-compliance
Lightning Source LLC
Chambersburg PA
CBHW020101020526
44112CB00032B/785